Managing Pastures and Cattle Under Coconuts

Westview Tropical Agriculture Series
Donald L. Plucknett, Series Editor

Managing Pastures and Cattle Under Coconuts
Donald L. Plucknett

Because of the long life of a coconut palm--sixty to eighty years--and the relatively wide spacing the plants require, every coconut grower faces the problem of how to manage the land beneath the palms. Many of the small-scale farmers who manage over 90 percent of the 6 million hectares of coconut palms in the world have learned that raising cattle or other livestock under the palms can be profitable, as well as an effective method of controlling weeds.

This book reviews current knowledge on this productive farming system, drawing on research results and experiences of successful farmers. Well illustrated with photographs from producing areas, the book includes information on the management of both natural (unimproved) and improved pastures.

Donald L. Plucknett, professor of agronomy at the University of Hawaii at Manoa, is currently working as Chief of the Agriculture and Rural Development Division in the Office of Technical Resources of the Asia Bureau in the Agency for International Development. He was chairman of the National Academy of Sciences Vegetable Farming Systems delegation to the PRC in 1977.

To my father, W. Donald Plucknett,
whose first love has always been farming
and the land

Managing Pastures and Cattle Under Coconuts

Donald L. Plucknett

Routledge
Taylor & Francis Group

LONDON AND NEW YORK

First published 1979 by Westview Press, Inc.

Published 2018 by Routledge
52 Vanderbilt Avenue, New York, NY 10017
2 Park Square, Milton Park, Abingdon, Oxon OX14 4RN

Routledge is an imprint of the Taylor & Francis Group, an informa business

Library of Congress Catalog Card Number: 79-5357

ISBN 13: 978-0-367-01768-2 (hbk)
ISBN 13: 978-0-367-16755-4 (pbk)

Contents

Tables

Figures

Foreword

This book provides important insights into a form of economic development that utilizes existing natural resources to increase productivity of the land, and thereby improves incomes for the large number of people (predominantly small holders) who own and manage coconut groves. Without displacing the present groves, the establishment and maintenance of nutritious pastures under the trees for providing milk and meat from grazing animals should provide an economic thrust for progress throughout the coconut growing areas of the world. The improved economic stability of such combined enterprises should benefit the parent nations as well as the basic producers on the land.

This type of application of modern technology may well be extended to other tree crops (such as the oil palm), particularly in the early years of establishing such groves, when no income is produced otherwise. Although greater managerial skills are needed to satisfy growth and production requirements of both trees and pastures, the need for additional inputs should be relatively small, and the productivity of both land and labor stand to benefit handsomely.

The concept of multicropping of lands occupied by tree crops warrants exploration as to the potential for producing selected food crops, as well as forages for livestock. This book is a significant contribution to the application of that concept.

Howard B. Sprague
Agricultural Consultant

Preface

Agriculturists have increasingly become aware that mono-cropping or single crop strategies alone are not satisfactory in meeting the food or income needs of small farmers in the tropics. This awareness has led to a growing interest in multiple cropping and in studying existing systems of farming. Today, many research institutions around the world are conducting research on cereal/food crop multiple cropping systems and on specific problems facing small farmers. Most of the research is devoted to combinations of annual crops, and little is known of multiple cropping in the many perennial tree or plantation crops of the tropics. This work constitutes an effort to analyze one such farming system, the coconut/pasture/cattle complex.

This book is the outgrowth of a number of events and experiences over the past seven years, all of which have served to convince me that the coconut/pasture/livestock farming system is far more important than many people are aware, and that its full potential is not being realized because of a general lack of information and understanding about the system itself. Originally, it began as a background review paper for the South Pacific Commission (SPC) Regional Seminar on Pastures and Cattle Under Coconuts, held in Alafua, Western Samoa on August 30-September 12, 1972. That review paper was based largely on published information from Sri Lanka (formerly Ceylon) and from tropical pasture research programs in Australia, Hawaii, and other research centers.

At the South Pacific Commission seminar, new

information and insights were gained through discussions and interactions with seminar participants from several South Pacific countries and territories. At that time it became clear that much more was known about cattle and pastures under coconuts than had been published, and that farmers and agriculturists in several countries had gained an impressive knowledge of this balanced farming system and its management requirements. It also became apparent at the seminar that someone should systematically gather the accumulated knowledge, published and unpublished, and then analyze and distill it into a form where it could be put to work on the fields and farms of coconut-growing countries. Such a state-of-the-art study was also needed to identify research and development needs for improvement of this farming system.

To complete this work required sustained library searches for obscure publications and references, field visits and surveys on coconut/cattle farms in several areas, visits to coconut research centers, and the full cooperation of a number of interested key individuals.

Visits (not exclusively for this purpose) to the Philippines, Western Samoa, Sri Lanka and Jamaica allowed me to study natural and improved pastures under coconuts over a wide geographic area. The writings of observant, innovative scientists and agriculturists provided access to accumulated practical knowledge in several countries. Much of the writing has been accomplished on airplanes, in airports, and while enroute to or from various developing countries on official travel for the Agency for International Development and the University of Hawaii.

I am indebted to many persons and institutions for providing information, assistance, and encouragement; not all of these can be named, but some stand out as especially deserving mention.

Leaders of the College of Tropical Agriculture, University of Hawaii, Honolulu; Dr. William R. Furtick, Dean of the College; Dr. C. Peairs Wilson, formerly Dean of the College; Dr. K. K. Otagaki, Director of International Agricultural Programs; and Dr. Wallace G. Sanford, former Chairman of the Department of Agronomy and Soil Science for supporting me in the work.

The South Pacific Commission and its former
Livestock Officer, Mr. Edwin I. Hugh, for making it
possible for me to participate in the SPC Regional
Seminar on Pastures and Cattle Under Coconuts, and
to the participants in that meeting (for a list of
participants, see Appendix A).

The Western Samoa Department of Agriculture,
William Meredith, Director, and Frank Moors, former
Deputy Director; the Western Samoa Trust Estates
Corporation (WSTEC); Morris Lee, General Manager,
and Willie Wong, Assistant General Manager; and
leaders and staff of the UNDP Special Project in
Western Samoa for making it possible for me to
visit and study cattle/coconut farms and research
programs.

To my colleagues, Moises de Guzman, Livestock
Officer of the ASPAC Food and Fertilizer Technolo-
gy Center, Taipei, Taiwan; staff members of the
Ceylon Coconut Research Institute; Dr. K.
Santhirasegaram, former agrostologist; D. E. F.
Ferdinandez, Agrostologist, and Dr. U. Pethiyagoda,
Director; Drs. G. Osborne, P. Whiteman, and L. Ross
Humphreys, University of Queensland, Brisbane,
Australia; Dr. Brian Robinson, formerly on second-
ment from the Ministry of Overseas Development,
United Kingdom, to the Department of Agriculture,
Fiji, and the Jamaica Coconut Industry Board,
R. A. Williams, Manager/Secretary, Kingston.

The Board of World Missions of the United
Methodist Church of the United States, and the
Samoan Methodist Land Development Incorporated for
travel support and practical experience gained.

Additionally, there are numerous other individuals, both producers and scientists, who have
assisted in one way or another to help to complete
this book. Some of their names appear in the list
of references, some are credited as giving personal
communications to me. To these and to those un-
named persons, I wish to express my personal
gratitude.

The manuscript was typed by Margaret Sung
White, Sylvia Lau, Kathleen Sullivan, Mrs. Guineve
Tjossem, Mrs. Barbara Trueheart, Ms. Vivian L. McBee

and Ms. Barbara Becker. Dr. Howard B. Sprague has reviewed and commented on most of the chapters.

It is my hope that this book will stimulate further development and improvement in coconut/pasture/livestock systems, and that it may help in at least a small way in providing more food and improved income for coconut farmers.

Donald L. Plucknett
Washington, D. C.

1
Introduction

There are probably more than 6 million hectares (ha) of coconut (Cocos nucifera L.) in the world; about 90 percent of this crop area is in Asia and Oceania (Figure 1.1). Major producers of copra and coconut oil are the Philippines, Indonesia, Sri Lanka (formerly Ceylon), Mexico, Malaysia, and the islands and territories of Oceania. The crop is grown mainly on islands, peninsulas, and along coasts, partly as a result of favorable growing conditions as well as easy access to shipping and transportation facilities.

Coconut may be the most extensively-grown tree crop in the tropics. Truly a wonderful plant, it supplies man with many products. The leaves provide thatch, building materials, and items for crafts; the trunks make good posts or lumber for native houses; the base of the trunk is used for making drums; the husks are used for fuel and to provide coir fiber for making sennit rope and twine; the nut shells are used to make charcoal, small bowls and utensils and for making jewelry and other craft items; the kernel is used for food, for copra and as a source of "coconut milk" and coconut oil; and the water is often drunk for refreshment as a cooling drink. Alcoholic drinks or "toddy" are made by tapping the immature flower spathes.

Although coconut is often thought of as a large-scale plantation crop, most of the world production comes from small farms. Because world copra prices fluctuate widely and are frequently low, coconut-growing areas often experience economic instability, leading to neglect and poor condition of the palms. The coconut industry would

1

benefit greatly from diversification to provide economic stability for producers.

One way to obtain more income on coconut farms is to practice mixed farming or intercropping (Santhirasegaram, 1966b, 1966d, 1966e; Ohler, 1969; Plucknett, 1972a, 1972b; Ferdinandez, 1973). This results in more intensive land use, stabilization of the economies of the countries or islands dependent upon coconut for export income, increased food production, and improved rural welfare. A quote from FAO (1966) states the case very well; "The expansion of coconut as part of a mixed farming system may well make the difference between living, and merely existing at subsistence level, for the large mass of small holders in the coconut growing countries of the world."

Coconut is often a neglected crop, and its management is commonly poor. This is especially serious when one considers that: (1) it usually occupies good lands, and (2) it is such a long-lived crop, from 60 to 80 years or more. Poor management leaves lands in a nearly idle and unproductive state, denying producers and their countries needed food production and export income. Under-utilization of land may be one of the most important reasons to consider mixed cropping on coconut farms.

Coconut growing areas are a logical target for more intensive land use. They usually are located near coasts with good access to shipping and marketing routes or near main market roads; therefore, livestock or crop products can be easily transported and marketed. Also, coconuts often occupy some of the best land resources of a country; they are usually planted in or near high rainfall coastal areas with good topography where more intensive land use is needed. Oram (1975) made a strong plea for fertilized pastures under coconut and oil palm (Elaeis guineensis) to obtain increased feed for livestock without tying up additional land.

When lightning damage, attack by disease or insects, droughts or other catastrophes kill palms (Figure 1.2), farmers are left in a dilemma. Faced with uneconomic palm stands and barren sections of fields, farmers must decide whether to cut down remaining productive palms, which constitute a significant long-term investment, and then replant,

2

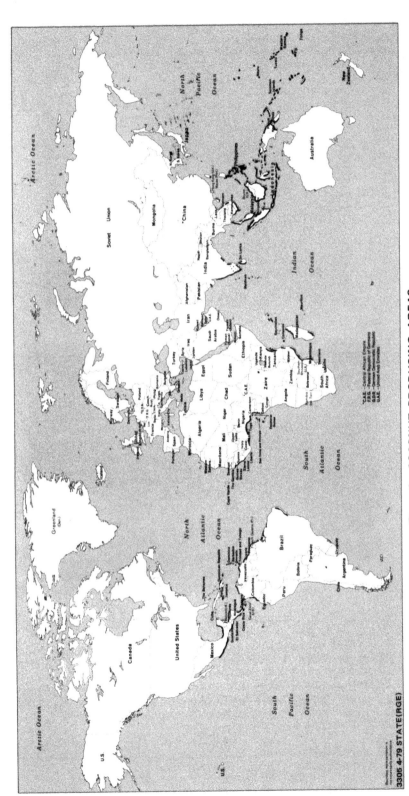

Figure 1. COCONUT–PRODUCING AREAS

Figure 1.2. Above: Aerial view of old coconut
groves in which natural pastures are grazed
for beef production. Note wide spacing
between some of the palms; some clearings and
open spaces have resulted from hurricane damage.
Below: Ground view of the pastures.

or attempt to continue coconut production and find other ways to increase farm income and food production. Intercropping with food crops or raising cattle on pastures under the palms are ideal solutions to increase production and income in damaged groves while rehabilitation or underplanting steps are being carried out. In sole coconut enterprises, badly damaged groves often must be abandoned, and farmers are left with reduced income or sustenance. With intercropping or livestock production under the palms, however, damaged groves can remain in production while producing needed income.

Better labor utilization is possible with mixed farming. Because most coconut producers are smallholders, this becomes of major importance. Mixed farming allows better use of family labor and provides profitable cover management of the groves instead of costly and less profitable weeding practices.

Smallholders also benefit from increased food production on their lands. In the case of coconut/ pasture/cattle enterprises, provision of milk and meat for the family may be an important consideration, in addition to possible sales of produce.

Local markets for milk or meat provide outlets for the animal products of coconut/pasture enterprises. Many tropical islands and developing countries are net importers of processed milk and animal products. With good transportation and close proximity to coastal cities, coconut/pasture farmers should be able to find ready markets for their products. For many countries, the need to export meat would be a prospect for the distant future, if ever. Farmers will, therefore, supply their family food needs first, then local market needs, and lastly will look for export markets.

Most coconut products, chiefly copra and coconut oil, are exported. For this reason, transportation becomes a major factor in coconut production. Many coconut growers depend upon river, coastal or inter-island shipping to consolidate their products at ports handling international shipping trade. The industry depends to a lesser extent on land transport, except for shorter hauls to local ports or processing centers. Some of the small islands of Oceania which are dependent upon copra for export income suffer greatly from deficiencies in their transportation systems; such deficiencies also affect markets for intercrops or animal products.

Mixed farming on coconut farms is not new. Chapter 4 will discuss past intercropping experiences as well as possibilities for the future. However, the coconut/pasture/cattle enterprise has not been fully discussed or evaluated, especially from the coconut/pasture management point of view, nor from the standpoint of a critical analysis of its value as a farming system for small farmers in developing countries.

This book will evaluate the considerations and steps to take for managing coconut/pasture/cattle farming systems. It will look at both natural and established pastures, but will emphasize the importance and advantages of establishing improved pastures under coconuts. It should be understood that these are systems of management, and that two or more products or enterprises are being produced on the same land. Therefore, both products -- the basic perennial crop (coconut) and the intercrops (annual food crops, semi-perennial crops or permanent crops, including pastures) -- must be managed in accordance with their individual and collective needs. To do otherwise could prove unwise or even unprofitable. This does not mean, however, that the resulting coconut/pasture system has to be complicated; but it does mean that consideration must be made for each of the combined enterprises.

Although this work will restrict itself to the coconut/pasture farming system, the tree crop/pasture system could be useful in other tree crops such as oil palm. Indeed, Hartley (1970) reported that livestock/pasture systems are employed in oil palm in Asia and Latin America, and that research on pastures under oil palm has been conducted in Nigeria and Sierra Leone. Other crops in which this system might work include cashew (Anacardium occidentale) and young rubber (Hevea brasiliensis).

ADVANTAGES OF COCONUT/CATTLE ENTERPRISES

Cattle and other livestock have been raised on coconut farms for a long time. Probably the main reasons for this practice in the past were: (1) to help keep down weed growth, (2) to provide food for

6

draft, dairy, beef, and other animals, and (3) to provide manure for coconuts. In recent years, however, many coconut-producing countries have come to realize that animal production on coconut farms can also raise farm income through sale of meat or milk, while at the same time reducing weed control costs by growing a dense forage cover under the palms. Related benefits include possible improvements in structure, fertility, or moisture-holding capacity of the soils.

Some obvious advantages of raising cattle under coconuts are: (1) increased farm income, (2) reduced weed competition and weed control costs, (3) better land use by intercropping in the open ground beneath widely-spaced palms, (4) increased food production, (5) to provide feed for draft animals, milk cows and beef animals, and (6) production of manure to improve soil fertility, soil structure, and moisture-holding capacity. A less obvious, and sometimes debated, advantage is higher coconut yield with good pasture management.

Disadvantages of the coconut/pasture system can include: (1) damage to palms by animals, (2) competition between palms and pastures for plant nutrients and moisture, (3) soil compaction by animals, (4) with over-grazing, soil erosion and loss of soil fertility may result, (5) greater capital requirements for the two enterprises, and (6) a requirement for greater management skills and knowledge.

IMPORTANCE AND STATUS OF COCONUT/PASTURE/CATTLE OPERATIONS

Coconut lands are used for grazing in many countries, but information on the status of these enterprises is difficult to obtain. Short descriptions of the industry in some countries may help to point out the present usefulness and potential for this farming system.

Asia

Asian countries lead the world in coconut production and crop area. For the period 1958-61, Asian production of copra and coconut oil was 78.4 per cent of world production (Espenshade, 1964). Present land area planted to coconut in Asia is about 6.1 million hectares, about 90 per cent of the estimated world crop area (Table 1.1).

Within Asia, Southeast Asian countries dominate in copra and oil production and in crop area. The Philippines and Indonesia together total more than 3.9 million hectares of coconuts; this is 58 per cent of world crop area.

Livestock are grazed under coconuts in most of the countries, and the potential for expanded production is great. For example, on the basis of an FAO (1971) estimate of 3.1 million ha of bearing coconuts in the Southeast Asian countries (from Burma to the Philippines, but excluding Papua New Guinea), Dr. Lito Calo (undated) estimated that more than 6 million cattle or domestic buffalo could be grown in the region without opening new lands for pasture.

Philippines

There are some 2.1 million hectares of coconuts in the Philippines (de Guzman, 1974), although other recent estimates have placed the figure at 1.8 million ha (Tanco, 1973). Cattles are grazed on about 440,000 ha (MacEvoy, 1974).

Smallholders with less than 2 ha comprise some 10 per cent of this crop area (Asian Development Bank, 1969); however, about 90 per cent of the industry is comprised of small farms from 0.1 to 20 ha in size (Cornelius, 1973). About 40 per cent of the coconut lands are in Mindanao, with another 20 per cent in the eastern Visayas. Most of the farms which graze livestock under coconuts are located on Mindanao (Barker and Nyberg, 1968). Livestock grazed under coconuts includes horses, water buffalo (carabao), and cattle.

Pastures under coconuts are mainly composed of a natural growth of weeds or cover crops. Carrying capacity of the pastures is about 1/2 to 1 animal per ha (Tanco, 1973). Currently, there is much interest in improving pastures under coconuts for beef production; such improved enterprises are popularly called "coco-beef" (MacEvoy, 1974). Improved grass/legume pastures can carry as high as two animals per ha, year round, on a rotational grazing basis without detrimental effects on the coconuts (Dr. Lito Calo, undated).

8

Barker and Nyberg (1968) surveyed 1,230 coconut farms in 12 provinces. Farm size was found to be an important factor in coconut/pasture/livestock enterprises. For example, 10.9 per cent of farmers with less than 2 hectares grazed livestock on their lands, while 52 per cent of farmers with over 49 hectares grazed livestock under coconut. A comparison was made between three types of coconut/beef cattle enterprises. These were:

(1) A coconut/beef enterprise of 150 hectares carrying one mature animal per ha; without improved pastures and without use of concentrate feeds and with little fencing. This enterprise was largely designed to keep down weeds. Calf crop was estimated at 55 per cent, death loss at 10 per cent, and replacement rate at 10 per cent. Thirty animals were sold each year, averaging 250 kg each. The major investment was livestock. For this enterprise the return on capital was 13 per cent.

(2) A coconut/beef enterprise of 150 hectares of which 100 hectares was improved pasture carrying three mature animals per hectare. Calf crop was estimated at 80 per cent, calves at weaning are fed for 260 days and marketed at 18 months weighing 330 kg. Rate of gain in feed lot -- 0.5 kg/head/day. Death loss -- 10 per cent, replacement rate - 10 per cent. This enterprise required considerable capital, nearly four times that for the traditional method, for pasture improvement, machinery and equipment and for fencing. Return on investment was 10.7 per cent.

(3) A coconut/beef cattle enterprise of 300 hectares (double that of 2), using the improved pastures and practices listed above for that operation. For this enterprise the return on investment was 13.9 per cent, reflecting largely economies of scale.

For a later study of the economics of beef production under coconut in the Philippines see de Guzman (1970).

Table 1.1
Estimated Crop Area of Coconut Across the World (in thousands of hectares)1/

ASIA	000 ha	OCEANIA	000 ha	AFRICA	000 ha
Philippines	2,100	Fiji	72	(EAST)	
Indonesia	1,870	Solomon Islands	40	Kenya	20
Papua New Guinea	249	Gilbert & Ellice Is.	8	Tanzania	41
India	906	Trust Terr. Pacific Is.	30	Seychelles	
Sri Lanka	445	Western Samoa	29	Islands	12
Malaysia	246	Kingdom of Tonga	24	Malagasy Re-	
Thailand	318	French Polynesia	60	public	15
Vietnam	44	New Hebrides		Mozambique	57
		Condominium	30		151
		Cook Islands	4		
				(WEST)	
				Ghana	14
				Nigeria	8
				Benin	12
				Togo	8 (?)
				Guinea-Bissau	6 (?)
				Ivory Coast	5 (?)
					53
Sub-total	6,178	Sub-total	297	Sub-total--Africa	204

Table 1.1 (con't)

WEST INDIES	000 ha
Islands, formerly	
Br. West Indies	32
Puerto Rico	4
Dominican Republic	27
Jamaica	16
Sub-total	79

CENTRAL AND SOUTH AMERICA	000 ha
Mexico	81
Central America	10
Guyana	14
Belize	1.6
Surinam	2
Sub-total	108.6

GRAND TOTAL WORLD-WIDE ---- 6,865,000 ha

1/ Reliable statistics on coconut hectarage are extremely difficult to obtain, for the plant is often grown as a village or interplanted crop, and figures are often based on palm counts as well as land area surveys. These data are based in part on figures presented by Child (1964), but have been updated using more recent estimates for several countries (Asian Development Bank, 1969; Selvadurai, 1968; Unpublished data of Dr. Arkoon Soothipan, Kasetsart University, Bangkok; Thailand; Philippine Council for Agric. Res., 1975; Agency for International Development, 1976).

Indonesia

There are more than 1.8 million hectares of coconuts in Indonesia (Asian Development Bank, 1969). The main production area is North Sulawesi, but other important production centers are the Moluccas, North Sumatra, and the Riau Archipelago (Asian Development Bank, 1969). Although current figures are not available, it appears that most of the production is in the hands of smallholders, for at the time of World War II only about 5 per cent of the crop area was devoted to plantations (Cornelius, 1973). Information on the coconut/pasture/cattle system is available for the island of Bali where cattle graze mostly weeds and shade tolerant plants of natural pastures on sandy soils in coastal areas where rainfall is lower and the human population is low (Steel, 1974; Steel and Humphreys, 1974). In higher rainfall areas where population density is great, intercropping with food crops is the predominant system in coconut lands. Cattle grazed under coconuts in Bali belong to the species, Bos banteng, which is indigenous to Southeast Asia.

Malaysia

Malaysia has an estimated 246,000 hectares of coconuts, of which 222,000 ha are in West Malaysia (Selvadurai, 1968). An Asian Development Bank report (1969) indicated that there were 202,000 ha in West Malaysia. Cattle and buffalo are grazed on natural pastures under coconut, but little published information is available. Natural pastures, if closely grazed, are composed mainly of Axonopus compressus, Chrysopogon aciculatus, and Desmodium triflorum (Verboom, 1968). Selvadurai (1968) conducted a detailed survey of small-scale coconut producers in West Malaysia; this survey is of great value in determining the production system used by small farmers. McCulloch (1968) reported that 87 per cent of coconut lands in Malaysia were managed by smallholders with 40 hectares or less.

Papua New Guinea

There are about 250,000 hectares of coconuts in Papua New Guinea. About one-third of the production is in the hands of smallholders (Cornelius, 1973). Many palms were damaged or neglected during

World War II. Some cattle are grazed on natural
pastures, mostly to control weeds under the palms
(Charles, 1959; Hill, 1969). Cacao (Theobroma
cacao) is the most important intercrop under the
palms.

Sri Lanka (Ceylon)

No country has done as much to dramatize and
rationalize the coconut/pasture/cattle farming system as Sri Lanka. The country has about 445,000 ha
of coconut, most of which is produced in the southwest and west coastal lowlands of the island in a
high rainfall area popularly known as the "Coconut
Triangle." About 75 per cent of the total coconut
land area is in the hands of smallholders with less
than 8 ha (Asian Development Bank, 1969). The coconut lands have long been important in production of
milk and meat; in 1968 the livestock population of
the Coconut Triangle included 337,000 cattle and
181,000 buffalo (Buvanendran, 1970).

Most of the pastures under the palms are used
for dairy cattle to produce milk for the Colombo
market, as a subsidiary enterprise with coconut.
Pastures are comprised mainly of shade tolerant
weeds and adventive species, although there is an
increasing area being planted to improved pastures.
Appadurai (1968) estimated that about 142,000 ha
could be developed to improved pastures under coconut. Recent government policies have been designed
to encourage expansion of improved pastures. Carrying capacity of natural pastures is about 1 animal per ha (Ferdinandez, 1975).

The Coconut Research Institute of Ceylon has
pioneered in pasture research under coconuts, and
its careful studies have been of real value elsewhere. One of its major achievements was the
recognition that Brachiaria miliiformis, named
"cori" grass is honor of the Institute, was an excellent pasture grass in lightly-shaded mixed
cropping situations.

India

India has some 906,000 ha of coconut, most of
which is located on the southwest coast of the sub-
continent (Asian Development Bank, 1969). Kerala
State is the most important growing area, with 60

13

per cent of the crop area and 70 per cent of the total production. Coconut is mostly a smallholder crop, with 90 per cent of the crop area devoted to farms of 2 ha or less (Asian Development Bank, 1969). Cattle and buffalo are grazed under the palms, probably on native pastures. Recently, research in improvement of fodder production and pastures under coconut has been initiated at the Central Plantation Crops Research Institute at Kayangulam (Sahasranaman and Menon, 1973). These studies emphasize systems for small farmers.

Oceania

The islands of the Pacific depend heavily on coconut for food, fiber, building and craft materials, and export income. These islands have about 4.3 per cent of the world coconut lands. Diversification in the coconut lands would be of great benefit to the island people.

Some of the best information on coconut/pasture/cattle systems, especially for small farmers, has been gathered from the South Pacific Regional Seminar on Raising Cattle Under Coconuts, sponsored by the South Pacific Commission and held in Western Samoa on August 30-September 11, 1972. Participants from several island countries and territories brought information from their jurisdictions, and fruitful discussions were held on various topics relating to the potentials and problems of this farming system. A very useful and interesting report of the seminar has been published (South Pacific Commission, 1972; Hugh, 1972b).

Fiji

There are about 72,000 ha of land planted to coconut, of which an estimated 3,000 ha are devoted to pasture (Satyabalan, 1972). While natural pastures are grazed, the government is also encouraging the establishment of improved pastures (Ranacou, 1972a). Carrying capacities are about 0.6 to 0.75 animals per ha. Fiji is self-supporting for fresh beef, but it imports canned beef. Coconut-growing islands are remote and are faced with transportation difficulties. About 60 per cent of coconut lands are owned by villagers.

14

Western Samoa

This small independent island country now has more than 29,000 ha of coconut, and is engaged in an active replanting program. Most of the communal village coconut lands are intercropped with food crops, but some private farmers, church missions, and the large publicly-owned plantations do graze cattle on natural pastures, and increasingly, on improved pastures, mostly of Brachiaria brizantha. Some communal villages are beginning to develop their own beef and dairy herds.

Palm spacing in Samoa averages 9 X 9 meters, in order to make intercropping possible. This spacing allows good pasture growth.

A highly successful program to introduce dairying to remote villages has been devised by Frank Moors, formerly Deputy Director of the Department of Agriculture. In this program, village Women's Committees (originally quasi-religious organizations now devoted to social welfare programs) purchase dairy heifers at low cost from the government. The Women's Committees have organized themselves to care for, feed, and milk the animals. They sell the fresh milk to children who drink it on the spot. The Department of Agriculture provides a mobile breeding service, in some cases leading the bull to remote villages not served by roads. Increases in both milk and beef have resulted. Some of the feed for these animals comes from natural pastures under coconuts.

Solomon Islands

There are about 40,000 ha of coconuts in the Solomon Islands, and about 14,000 cattle graze on some 24,000 ha of these lands. The cattle are used mostly as weeders on native pastures. The government would like to increase cattle numbers and to concentrate and narrow its efforts on lands best suited for improved pastures (Dr. Ian Freeman, Pers. Comm., 1972). Carrying capacity of pastures is 2 animals per ha for Brachiaria mutica and 0.5 animals per ha for Pennisetum polystachyon.

New Hebrides

There are about 30,000 ha of coconuts in these islands. Most of the palms are old or senile, and stands are very poor, due in part to hurricane damage. Pastures under the palms are mostly natural and cattle are used as weeders (Weightman, 1977). There are many large plantations. The New Hebrides are now self-sufficient for beef and are canning and freezing some of the surplus for export to New Caledonia. A 600 hectare coconut farm can produce 50,000 kg of meat per year (Dr. R. Valin, Pers. Comm., 1972).

Tonga

This small island kingdom started its cattle production program by tethering animals to trees. The present land tenure system calls for each male child to be given 3.3 ha of land at maturity, and intensification of agriculture is necessary, for the land grant to individuals was once much larger. Therefore, there is interest in planting improved pastures and in fencing. Females are bred on a "honeymoon" scheme at the government farm (Siaosi Moengangongo, Pers. Comm., 1972). Over 50 per cent of the agricultural land of Tonga is planted to food crops, and most intercropping is practiced under coconuts.

Niue

This tiny island is experimenting with an approach to cattle production under coconuts on family or communal lands. The government encouraged eight families to commit 40 to 60 hectares each to a management board which then plants improved pastures and manages the cattle, while paying land rental to the owning families who collect and use or sell the nuts (Richard Lucas, Pers. Comm., 1972). Carrying capacity reaches or exceeds 2 animals per hectare. There are about 4,000 ha of coconut on the island, about half of which is presumed to be suitable for coconut/pasture/cattle.

AFRICA

Coconut is not a major crop in most African countries, but it is grown along the coasts of both East and West Africa. Although firm statistics are difficult to obtain, the continent probably has about 3.0 per cent of the world coconut hectarage. Besides countries for which statistics are given in Table 1.1, other countries or territories which grow the crop include Sao Tome and Principe and the

16

Comoro Islands (Van Chi-Bonnardel, 1973).

Tanzania

This East African country has about 41,000 ha
of coconut. Childs and Groom (1964) reported that
grazing under coconuts was not normal practice because of fragmented land holdings. These workers
implemented a pilot project to improve milk production under the palms, and to provide a working
arrangement between livestock owners and coconut
farmers which provided grazing lands for cattle
while helping to control understory weed growth in
coconut.

Mozambique

There are some 57,000 ha of coconuts in Mozambique. Child (1955, 1964) described the importance
of large cattle herds in maintenance of plantations.
Cattle manure is collected for use on the palms.
An unusual system of planting in clusters of four
closely-spaced palms to provide open areas for
pasture production has been tested at a large plantation at Gurae. This system is called "bouquet"
planting.

THE AMERICAS

As in Africa, coconut is not a major crop in
most American countries. Mexico has the most land
devoted to the crop, 81,000 hectares (Table 1.1).
Despite generally poor statistics, it appears that
the Caribbean Islands have about 40 per cent of the
total hectarage in the region. Countries or territories which grow coconuts but which are not listed
in Table 1.1 include; Honduras, Nicaragua, Panama,
El Salvador, Ecuador and Colombia.

Jamaica

There are about 15,000 ha of coconuts in
Jamaica. Lethal yellowing disease has caused
death of many palms, and the resulting thin stands
have caused farmers to practice considerable intercropping, especially with banana (Figure 1.3).
Cattle are grazed on natural pastures in many parts
of the island; usually poor coconut and poor pasture are grown together (Ann. Rep., Coconut Industry Board, 1964). In 1961 the Research Department

17

of the Coconut Industry Board initiated research
in managing improved pastures. These experiments
were terminated in 1971 (Ann. Rep., Coconut Industry Board, 1961 - 1971).

Dominican Republic

There are some 27,000 ha of coconuts in the
Dominican Republic (Agency for International Development, 1976). The crop is grown on the north
coast of the island, especially along the northeast
coast. Little information is available on the industry.

Figure 1.3. Bananas grown under coconuts in Jamaica where lethal yellowing
has killed many palms and decimated
stands.

2
Cultural Requirements
of Coconuts

If an intercrop is to be grown with a long-term crop such as coconut, the second crop must fit into the management system of the permanent crop. For that reason, it is necessary to consider the management requirements for coconut, especially as they relate to multiple use of coconut groves or plantations. This chapter cannot be more than an overview of some factors of coconut management; for a more complete discussion, there are several books and bulletins on coconuts (Ferguson, 1907; Copeland, 1914; Munro and Brown, 1916; Patel, 1938; Eden, 1953; Menon and Pandalai, 1958; Child, 1964; Piggott, 1964; Migvar, 1965; Fremond, Ziller and de Lamothe, 1966; Lambert, 1970; Sanchez Potes and Mena, 1972; Markose, 1973; Philippine Council for Agricultural Research, 1975; Anonymous, Undated). For discussions of other specific management practices in coconut, see Chapter 3 (cover crops, weed control, prevention of soil erosion, and fire); Chapter 4 (competition for water, nutrients); and Chapter 6 (tillage and the coconut root system).

DISTRIBUTION

The coconut palm is grown mostly on islands and coasts of the tropics and subtropics, but is only important as a crop within a widely-variable belt which is broadly outlined by the Tropics of Cancer and Capricorn. It cannot stand cold weather, and although palms may survive outside the Tropics, they usually do not produce fruit there. Mainly a crop of coasts, islands and peninsulas, coconuts can be grown at elevations of more than 600 m near the equator, but near the Tropics of Cancer or Capricorn are found mostly below 160 m.

THE PLANT ITSELF

Coconuts take six or seven years to reach production from transplanting (Figure 2.1). The trunk does not form until about 4 or 5 years after planting, although varieties do vary in length of time to flowering. The palms are monoecious, bearing both male and female flowers in the same inflorescence, and cross pollination is common. Most of the fertilized young fruits (buttons) fall long before maturity. It takes about 12 months for the remaining fruits to mature.

The roots are coarse and fibrous with few branches. Their depth and lateral spread vary greatly with soil type. In sandy soils the diameter of spread may be as much as 10 m or so. The roots will not develop in waterlogged soils. Most roots are found in the top 1 m of soil. For further information on coconut roots see Chapter 6.

Each coconut crop is the outcome of several events over a 3 1/2 year period (Abeywardena and Fernando, 1963). Krishna Marar and Pandalai (1957) pointed out that, "The primordium of the inflorescence is formed about 32 months prior to its opening, that of the spikes about 15 months before and of female flowers about 12 months before." It takes about 12 months after the spathe opens for the female flowers to develop into ripe nuts.

The components of yield in coconut are:

-- number of female flowers formed about 2 years before the crop matures,

-- fruit set - determined by number of female flowers which are pollinated; this occurs about a month after the spathe opens,

-- some of the young fertilized nuts fall from the spike, the extent of such "button shedding" is important in potential yield (Mathew, 1965),

-- number of filled nuts at maturity, and

-- copra or kernel weight per nut.

It can be seen from this listing of events concerned with fruiting that growing conditions over a

42 month period, from first initiation of the inflorescence to ripe nuts, can greatly influence yield.

In studies of the relative importance of crop yield components on final yield, Abeywardena and Fernando (1963) found that percent fruit set following pollination controlled about 50% of the yield fluctuations, and female flowers per bunch size controlled 20%. Premature fall of nuts and empty nuts did not influence more than 2 or 3% of the crop yield. These results show that climatic or other factors which cause losses or gains in per cent fruit set or in the number of female flowers per bunch will significantly affect yields. A palm produces about one new bunch per month (Figure 2.2), thus a tree will usually carry about 12 bunches approximately evenly spaced in their development stages (Abeywardena, 1955).

Coconut palms have an economic life of about 60 years, although they may still continue to produce for as much as 80 years or more.

RAINFALL AND YIELD

Coconut grows best in areas receiving 1250 to about 2500 mm of well-distributed rainfall. Below 1250 mm, unless supplemental water from irrigation or seepage of ground water from higher areas results, periodic moisture stress will occur, resulting in decreased production and crop injury. Above 2500 mm, unless soils are naturally well-drained or special drainage provisions are provided, the crop may suffer from excess soil moisture. The palms cannot stand waterlogged soil conditions. Coconut does respond to irrigation in dry areas (Wijewardene, 1957).

Rainfall is the single most important factor in coconut yield; however, the relationship is not simple nor easy to predict (Abeywardena, 1963, 1968; Lakshmanachar, 1963). Because flowers and bunches are being produced each month, and it takes about 42 months from initiation of the individual inflorescence to full maturity of nuts, low rainfall at any stage could seriously reduce yield. While it would be highly desirable to predict the exact effects of low rainfall at a given time, the problem is much too complex to predict satisfactorily (Abeywardena, 1971). Attempts to relate

Figure 2.1. A seedling nursery in the
Philippines.

Figure 2.2. Bunches of coconuts on a young
palm in Sri Lanka. Each bunch carries about
10 to 12 nuts.

yields to rainfall patterns have had limited value. Abeywardena (1955) found that yields were determined mainly by rainfall during the previous year and the first 3 months of the harvest year. Also distribution of rainfall was considered to be more important than total rainfall, because concentrated rain in a short time runs off and is of little value. The critical rainfall period during the year can be divided into sub-periods which can be used to study effects of rainfall on yield with fairly good results, improving yield predictions over the previous system of predicting yield on the basis of rainfall during the previous year.

Drought effects in coconuts include the following symptoms: drooping of leaves, breaking of petioles, and heavy shedding of immature nuts (Mathew, 1964). In prolonged droughts palms may die. A report on marginal lands in Sri Lanka (Anonymous, 1951) described the consequences of a severe drought in which an estimated one million palms died. Rainfall in the stricken area was about 1000-1250 mm per year.

Under severe waterlogging, palms may suffer slow growth and even death. Symptoms may include tapering of the trunk (sometimes referred to as "pencil point"), yellowing of leaves, and reduction in size and vigor of the crown. Provision of field drains and lowering of the water table can help to overcome this problem.

SOILS

Coconuts grow on a wide range of soils, from coastal sands to very heavy clays. Probably the best soils are well-drained alluvial deposits along rivers or estuaries. Sandy soils, oxisols, and ultisols are well-drained but somewhat infertile, and will require fertilization. Heavy clays can present serious physical problems and may require special management practices such as drainage, and plowing in of green manures or burial of husks to improve physical properties.

The palms can tolerate a wide range of soil acidity, from pH 5.0 or so for oxisols or ultisols to over 8.0 for coral sands. Most soil nutrient deficiency problems can be overcome by use of fertilizers, and animal manures (Anonymous, 1929, 1936).

23

Soil Management

Because coconut is a long-lived perennial crop, soil management practices relate mostly to soil and water conservation steps, fertilizer application and incorporation, and cultivation of intercrops. Many soil management practices suitable for coconut grown in monoculture may not be suitable for coconut/pasture/livestock systems.

Water Conservation. Because water is so important in coconut production, many steps have been taken to conserve rainfall and soil moisture. Common practices include: (1) weeding and clearing areas around palms prior to the rainy season, (2) mulching the area around each palm with husks or other debris, (3) bunding or ditching to control water, (4) cultivation to control weeds, (5) deep tillage or subsoiling to overcome soil compaction and to increase the soil moisture-holding capacity, (6) burial or incorporation of husks and other organic materials (including green manures) to increase the moisture-holding capacity of the soil, and (7) control of wide-scale burning of coconut lands. Cross bunds are constructed in Kerala, India to retain moisture in level or gently-sloping lands (Anonymous, 1971a). In steeper lands, crescent-shaped basins are constructed on the down-hill side of each palm and filled with soil or husks.

Drainage. To prevent damage to coconuts by water-logging, many farmers dig ditches or field drains to carry away unwanted water. To be successful over time, field drains require periodic maintenance (Agcoili, 1973).

Soil Conservation. Soil losses in coconuts that are overgrazed, mismanaged, or burned frequently can be severe (Gorrie, 1950). Bunding, terracing, planting of cover crops, and judiciously-timed cultivation can assist in reducing soil erosion. Of these practices, probably cover cropping may be the most successful and suitable (Felizardo, 1973). Well-managed pastures can be a major factor in preventing soil erosion.

Managing Sandy Soils. Sandy soils are often used for coconuts, and because of the low water and nutrient holding capacity, coconuts often do not yield well. Management steps to conserve soil moisture include planting of cover crops that are

24

less competitive for moisture during the dry season, incorporation of husks and organic materials to improve moisture-holding capacity, and improved nutrition of palms (Anderson, 1967).

Pests and Diseases

It is not possible to cover all pest and disease problems in this publication. However, it should be useful to discuss some problems which are pertinent to pasture use in coconuts. For details on major pests of coconut see Lever (1969).

Probably the most serious pest of coconut in Asia and the Pacific is rhinoceros beetle (Oryctes rhinoceros), which occurs from India, through Southeast Asia, to Indonesia and New Guinea, to Micronesia and southern Polynesia, and Melanesia (Catley, 1969). The adult beetle attacks the crown of the plant, boring into the crown and damaging the young leaves before they unfold. Severe attack may injure the growing point and kill the tree. Larva (grubs) of the beetle breed in refuse such as old coconut logs or stumps, dead palms, piles of rubbish or compost and animal manure, and dung heaps. Groves which are weedy, trashy, or otherwise poorly cared for or neglected contain many potential breeding sites for the beetles.

Control measures for rhinoceros beetle start with trash removal and burning, and clearing out of undergrowth and dead plant materials. Because the beetles can breed in animal dung, dung beetles have been considered as a means to remove dung as a potential breeding site, where grazing of cattle under coconut is practiced.

Aside from field sanitation, currently accepted control practices for rhinoceros beetle include use of a virus for biological control and trapping of adult beetles using attractants (Bedford, 1973).

Rats cause damage in palms, especially to young green nuts. Baiting and poisoning, and banding palm trunks with aluminum strips have been fairly effective. Keeping weed or understory growth low and under control can help to reduce habitats for rats; this, of course, is one advantage of closely-grazed, well managed pastures.

Diseases which cause palm death or gaps in

stand are important in coconut/pasture systems, for they may provide a special stimulus to find ways to increase farm income when such calamities strike. The most serious diseases are probably "cadang-cadang" disease in the Philippines (Philippine Council for Agricultural and Resources Research, 1975), "Lethal yellowing" in the Caribbean Islands (Romney, 1972), and a serious wilt disease in the coastal areas of Tamil Nadu, India (Natarajan, 1975).

HARVESTING

Nut collection is usually done by three basic methods; (1) gathering fallen ripe nuts from the ground beneath the palms (Figures 2.3, 2.4, 2.5), (2) pickers climb the palms and pick mature nuts before they fall (Figure 2.6), or (3) nuts are picked from the palms by men on the ground using long poles equipped with knives or saws to sever the bunches from palms. Following collection, the nuts are transported to the copra cutting and drying shed (Figures 2.4, 2.5, 2.7). It is generally agreed that best quality copra is obtained from nuts allowed to fall from palms at maturity. However, this is often not practiced and is even discouraged because fallen nuts may not be easily found in dense or tall understory vegetation and if not collected at frequent intervals may be damaged or spoiled because of sprouting or rat damage.

Harvesting for copra in the Pacific Islands consists of gathering nuts from the ground after they have fallen from the palms at maturity (Figures 2.3, 2.4, 2.5). Nut gathering must be done frequently to ensure high quality copra and to prevent the nuts from decaying or sprouting on the ground. Gathering schedules vary, but most producers make at least one round per month. For palms producing 50 to 60 nuts per year, this means that 3 or 4 nuts should be lying beneath each palm at each harvest round.

Nut collection is difficult if the ground is covered with dense or tall vegetation or weeds. In neglected plantations, production is reduced each year because nuts are not found and are lost through decay or sprouting.

Weed control by hand slashing or mowing helps to expose fallen nuts, but both may damage nuts

Figure 2.3. Nuts ready for collection in a natural _Axonopus compressus_/_Mimosa pudica_ pasture. Nuts are easily visible and gathering will be easy.

Figure 2.4. Nuts being collected by a gathering boy and a donkey equipped with a metal basket. Nuts are speared with a stick in which a headless nail has been driven, and sharpened to penetrate the husk for easy pick up. Western Samoan Trust Estates Corporation.

Figure 2.5. Samoan man collecting nuts in baskets woven from coconut leaves. Fallen nuts are collected and piled near field roads where small carts are used to haul them to the copra cutting and drying shed.

Figure 2.6. Coconut picker in Sri Lanka. These men climb the trees and cut or break off the nuts and drop them to the ground. They are paid according to the number of nuts harvested.

Figure 2.7. Some harvesting steps.
Top left: collected nuts awaiting cutting
for copra, Western Samoa: Top right: husking
nuts in Western Samoa; Bottom left: copra
cutting and drying shed and piled husks, Sri
Lanka; Bottom right: copra drying in the sun
on trays equipped with rollers and tracks so
that the trays may be pushed under cover in
case of rain.

unless the weeds are cut high, i.e., above the height of the nuts on the ground. This means that the nuts are still not easily seen from a distance.

Close grazing by animals helps to expose the nuts and make them visible for collection. In many coconut/cattle operations, nut collection follows right after the last grazing period. In the Solomon Islands in the South Pacific, nut collections have increased by as much as 20 per cent when cattle have been used to graze down weeds.

POTENTIAL YIELDS OF COCONUTS.

In considering pasture or intercropping under coconuts it is necessary to determine potential yields of both enterprises in order to have solid information on which to base management decisions. Of course, coconut yields will vary tremendously depending upon soil, rainfall, spacing, crop condition and other factors; however, the question of how much production can be expected from well-grown, well-managed trees must be answered. Lefort (1956) listed normal annual nut production figures for coconut as: 20 nuts at 8 years of age, 40 at 12 years, 60 at 16 years and 100 at 20 years.

What should be the yield goal for coconuts? Fremond, et al. (1966) gave a figure of 3000 kg of copra per ha per year for mature palms as a production goal, based on experimental results from the Ivory Coast. This yield was based on 140 palms per ha yielding 100 nuts per palm and requiring 5.6 nuts to produce 1 kg of copra.

Rockwood (1953) described a 3.23 ha commercial planting in Sri Lanka in which triangular spacing with 178 palms per ha was used. Planted in 1927, the field yielded just over 80 nuts per palm in 1952 at 25 years of age. The target yield for the field was 14,820 nuts per ha (83 nuts per palm) per year. Assuming 5600 nuts per metric ton (1000 kg) of copra, this would be equivalent to 2657 kg of copra per ha per year.

If these values are accepted as approximating a reasonable goal, how do current yields compare? Fremond, et al. (1966) presented some estimates of yield for various countries or areas. Estimated copra yields included the following values:

atolls of Polynesia - 200 to 290 kg per ha; India, Brazil, New Hebrides, Thailand - 390 to 490 kg per ha; Sri Lanka, Indonesia, Mozambique - 590 to 695 kg per ha; Ivory Coast, Solomon Islands - 795 to 880 kg per ha; Polynesian High Islands - 1000 to 1100 kg per ha; Philippines - over 1100 kg per ha.

From these and other yield reports it is probably safe to set a yield goal of about 11,000 nuts per ha per year, which should result in 2200 kg of copra per ha per year. Of course, not everyone will obtain this yield, but it is a very worthwhile target to aim for.

NUTRITIONAL REQUIREMENTS OF COCONUTS

In any consideration of intercropping in coconuts, it must be clearly recognized that two crops are being produced on the same land and that both crops must have adequate nutrition to remain productive. Indeed, such a system requires that both crops be fertilized in such a way that their full nutritional requirements are met. For that reason it is probably best to discuss nutrition of coconuts separately from nutrition of pasture species.

Nutritional Requirements of Young Palms

Coconut nutritional needs differ with age (Buchanan, 1966). Young palms respond readily to applications of N, P, and K, and especially to N and P. Nitrogen deficiency results in yellowing of plants, while K deficiency results in necrotic foliage tips. In young palms N is probably more important than P and K (Mathew and Ramadasan, 1964; Anonymous, 1967b). Phosphorus increases leaf collar girth and number of leaves, while potassium increases collar girth.

Fertilizing young palms with a complete NPK mixture is essential for young seedlings in underplanted coconut as well as in new areas (Salgado, 1951c; Nethsinghe, 1963; Sumbak, 1972). Complete NPK fertilizers improve growth, promote early bearing, and result in higher yields. In a newly planted area in Sri Lanka, 90% of NPK fertilized palms were bearing in eight years, while only 50% of unfertilized palms were bearing (Nethsinghe, 1963). At 13 years, fertilized palms yielded 76 nuts per tree while unfertilized palms yielded 40

31

nuts per tree. In underplanted palms, NPK-fertil-
ized trees yielded 62 nuts per tree at 15 years,
those without fertilizer yielded 30 nuts.

Palms that are not fertilized when young do
not reach the yield potential of fertilized trees,
even if fertilizers are applied later in the life
of the crop. In Sri Lanka, discovery of magnesium
deficiency in gravelly or light sandy soils has led
to the recommendation that dolomite be applied in
all planting holes (Nethsinghe, 1961).

Recommendations for young palms in Sri Lanka
are as follows, using an 8-5-8 mixture of N, P, and
K[1/] (Nethsinghe, 1963).

Time after transplanting	New clearings	Underplanted or replanted fields
	kg of mixture	kg of mixture
6 months	---	0.4
1 year	0.4	0.4
1 1/2 years	0.4	0.6
2 years	0.4	0.6
2 1/2 years	0.6	0.8
3 years	0.6	0.8
3 1/2 years	0.8	1.0
4 years	0.8	1.0

After the fourth year the mixture is changed
in order to increase K for the palms. The recommended mixture until bearing starts is a 7-2-14
mixture of N-P-K (not $N-P_2O_5-K_2O$), which is applied
biannually at 1 kg per tree on new plantings and at
1.2 kg per tree on replanted fields.

Shanmugan (1972) gave recommendations for fertilizing young palms in India. These are summarized in Table 2.1.

Nutritional Requirements of Mature Palms

One may ask how much nutrients are removed by
coconut during a year of full production. Values
reported have ranged from a low of 20 kg N, 2.2 kg
P, and 35 kg K, to 90 kg N, 18 kg P and 110 kg K

[1/] Please note that NPK values in this book are expressed as the pure elements, not P_2O_5 and K_2O.

Table 2.1. A suggested fertilizer schedule for young palms, per year (Shanmugam, 1972).

Age of Palm	Irrigated			Rainfed		
year	g/tree			g/tree		
1	60N;	20P;	120K	60N;	20P;	120K
2	120N;	40P;	250K	120N;	40P;	250K
3	250N;	60P;	500K	180N;	50P;	370K
4	360N;	100P;	750K	250N;	70P;	500K
5	500N;	140P;	1000K	310N;	90P;	630K
6	500N;	140P;	1000K	360N;	100P;	730K
7	500N;	140P;	1000K	410N;	120P;	860K

per ha per year. Pillai and Davis (1963) reported the following removal values for 5 nutrients from a sandy soil of average fertility in western India with 173 palms per ha, each palm producing 40 nuts per year: 95 kg N, 20 kg P, 110 kg K, 86 kg calcium and 34 kg magnesium.

The NPK requirements of adult palms are difficult to state directly, for they vary so much depending upon soil, spacing, climate and other factors. However, some general guidelines can be given (Anonymous, 1960). Nathanael (1967) presented fertilizer recommendations for various conditions in Sri Lanka (Table 2.2). The phosphorus recommendation for poor soils would appear to be quite low. Thampan (1970) recommended the following annual fertilizer application for adult palms in Kerala, India; 500 g N, 150 g P, and 1000 g K per palm.

Nitrogen has a significant effect on both copra and nut yields (Santhirasegaram, 1964) (Table 2.3).

For many older palms K may become a limiting nutrient. This is especially true on soils derived from coral (Charles, 1964) or in areas with high rainfall and with intensive soil weathering and leaching. In the wet zone of Sri Lanka magnesium deficiency has been observed (Nethsinghe, 1961). In order to overcome this problem, it was recommended that 0.68 kg of crushed dolomite be applied per palm in the high rainfall areas as a preventive measure.

Critical levels of plant nutrients in coconuts were presented by Fremond, et al. (1966). These values were based on leaf number 14, counting from the youngest leaf downwards, and are expressed as percentage of dry matter. The critical values are as follows:

nitrogen	1.8 to 2.0 per cent
phosphorus	0.12 per cent
potassium	0.8 to 1.0
calcium	0.5 per cent
magnesium	0.3 per cent
iron	50 ppm
manganese	60 ppm

Table 2.2. Fertilizer recommendations for adult
            coconuts in Sri Lanka (after Nathanael,
            1967).

Soil fertility condition	Nutrient required per tree		
	N	P	K
	g	g	g
good	360	110	450
fair	410	110	560
poor	450	110	680

Table 2.3 Effect of rates of nitrogen per palm on
           nut yield, copra yield per hectare, and
           copra yield per nut (Santhirasegaram,
           1964).

Nitrogen per palm[1) grams	number of nuts/ha	Copra	Copra yield per nut
g		kg/ha	g
330	6980	1630	235
650	7300	1730	234
1310	8180	1880	196

[1) biannual application

35

Sulfur Deficiency

Southern (1967a, 1967b, 1967c) described sulfur deficiency of coconuts in Papua New Guinea. This condition causes yellow or orange colored leaves, with necrosis following, resulting in death of leaflets and leaf tips. Arching of leaves commonly occurred. In older palms the leaves tend to bend above their normal abscission point, causing dead leaves to hang downward in a skirt-like pattern around the trunk. In severe cases few live leaves are seen, and these are usually small and stunted. Few nuts are produced, and these yield normal-appearing kernels, but upon drying result in rubbery copra. The problem can be overcome by treating with 900 g of elemental sulfur for mature palms, less for younger palms. Biannual applications of S may be necessary. Soils on which the condition occurs include those derived from young volcanic and alluvial materials. It should be pointed out that intercropping would increase the severity of untreated sulfur deficiency.

Chemical analysis of the "water" of the nuts can aid in detecting S deficiency (Southern, 1967b). Deficient palms yielded nuts in which S content of the water was below 8 ppm, while S-treated palms yielded nuts with water in which S levels were between 10 and 60 ppm.

Minor Elements

Coconuts on coral atolls may experience deficiencies of a number of elements including iron, manganese, zinc, copper, and other elements which become less available to plants because of the high pH (alkaline reaction) of the soils. Trunk injection of iron sulfate and manganese sulfate to overcome these problems was used by Thomson (1967) and by Southern and Dick (1967) with good results. Southern and Dick (1967) described the injection process. Six holes, 20 mm in diameter and 10 cm deep, were bored at equal distances around the base of the palm. Solid salts of $MnSO_4$ (5cc), $FeSO_4$ (5cc), $ZnSO_4$ (5cc), $CuSO_4$ (5cc) borax (5cc) and sodium molybdate (1cc) were forced into the holes with a plastic syringe and sealed with cotton and a sealing compound.

LOCAL FERTILIZERS

Local fertilizer sources (Anonymous, 1967c) may be used to help overcome soil fertility deficiencies. Some of these may include wood ashes, husk and shell ash from copra dryers, ashes from burning of stumps, trunks, trash, etc., resulting from field sanitation measures to prevent disease and rhinoceros beetle breeding sites. Other local fertilizers could include; animal manures, green manures, compost, household refuse, or even local phosphate or potash materials.

FERTILIZER PLACEMENT AND APPLICATION METHODS

Methods of fertilizer placement have been discussed for many years. Recommendations have varied from application in circular trenches around each palm to broadcasting over the entire surface (Nathanael, 1967). Most recommendations have included some form of fertilizer incorporation into the soil by hand hoeing (Figure 2.8), discing, or even burial. In radioactive P tracer studies in Sri Lanka, Nethsinghe (1966) found that most of the absorptive surfaces of coconuts were in a circle about 1.7 m in radius around the bole of the trees. This circle was judged to be the best area for application of fertilizers for adult palms. In later studies using labelled phosphorus Balakrishnamurti (1969) found that P uptake was most efficient at a depth of 10cm and at a distance of 0.5 m from the tree.

In a coconut/cattle enterprise, tillage to incorporate fertilizers would be costly and might cause damage to the pasture. For that reason broadcast applications will probably be best for both the coconuts and pure grass pastures. Work in Sri Lanka indicates that broadcasting of fertilizer may be superior to fertilizer burial. For legume-based pastures, broadcast applications of P and K should benefit both the legumes and the coconuts; however, broadcast N applications could create problems for the pasture legumes. Therefore, when pasture legumes are used, NPK fertilizers should be broadcast within a 3 m diameter circle around each palm, and fertilizers for the grass-legume mixture should be broadcast between the palm rows.

SPACING OF COCONUTS

Palm spacings used will greatly affect the potential for pasture production under coconut (Whitehead and Smith, 1968). Wider spacing will allow more forage growth, while close spacing will cause shading of the grasses and legumes. Shading effects will be influenced by both tree density and planting pattern.

Tree densities range widely; from as few as 49 trees per ha (spacing 13 x 13 m), 136 trees per ha (8.4 x 8.4 m), 158 trees per ha (7.8 x 7.8 m), to as many as 267 trees per ha (6 x 6 m) or more (Figure 2.9). As is common in most crops, yield per plant is greater at low plant populations and low at high populations, while yield per acre is highest -- up to a point -- at higher plant populations. Most countries recommend about 148 to 197 palms per ha.

Tree planting patterns can also affect shading and forage growth. Square patterns are probably used most frequently, but triangular or rectangular patterns are also used. A pattern called "quincunx" is designed to plant young trees in old plantations by adding one young tree in the center of each square of old trees (Ganarajah, 1954). Triangular plantings allow more trees per ha with more even distribution of trees in the field, resulting in an interlocking leaf canopy and more effective interception of sunlight. To determine the number of trees resulting from use of a triangular design, multiply the number of trees obtained using the same spacing in a square design by a factor of 1.155 (Abeywardana, 1954).

Table 2.4 shows the number of trees per ha possible under both square and triangular systems. This table should be of benefit in planning of new stands of palms.

UNDERPLANTING IN OLD COCONUTS

Should old, senile, low-producing coconut plantations be taken out entirely or should they be underplanted with young trees? This subject has been discussed for many years. Nair (1964) pointed out that, if we assume 70 years constitutes the normal productive life of a coconut palm, then it will be necessary to replant 1/70 of coconut lands

Figure 2.8. Man hoeing the "manure circle" around a palm in Sri Lanka. It is in this circle that manure or fertilizers are broadcast and incorporated into the soil.

Figure 2.9. A badly overcrowded grove in which copra yields will be low, and shading is too great to allow good growth of intercrop or pasture plants. Palm density in this field may be as high as 250 per ha or more.

39

Table 2.4. The number of coconut trees per hectare in relation to planting distance and planting pattern (after Sahasranaman, 1963).

Planting distance, in meters	Number of palms per hectare	
	Square Planting System	Triangular Planting System
6.6	222	257
6.9	202	232
7.2	188	215
7.5	173	198
7.8	158	183
8.3	148	170
8.6	138	158
8.9	128	148
9.2	118	138

each year just to stay even. While detailed discussion of underplanting is outside the scope of this book, decisions regarding underplanting (Figures 2.10, 2.11, 2.12) do have considerable bearing on the use of coconut lands for pasture and cattle production.

Many older coconut plantations are stocked with cattle. Underplanting of such lands could result in significant management problems including; fencing or protection methods for the young palms, weed control during the early growth of the young seedlings, provision for feed for cattle displaced from their former pastures in the old coconut fields, and methods of thinning of old palms and establishment of seedlings.

Complete clearing of old stands is one alternative. This method usually results in earlier and heavier yields for young trees because of lack of competition from old palms. It also may be most suitable for pasture use since the palms grow more rapidly and would require protection over a shorter period of time than would result from gradual thinning or late thinning. With complete clearing and replanting, establishment of new pasture species would probably be easier. However, if the old coconut lands contained improved pastures, some loss in use of the pastures could result with complete clearing unless the planting design used allows for suitable fencing at low cost.

Liyanage (1963a, 1963b) studied complete clearing, gradual thinning and late thinning of old plantations in Sri Lanka. Complete clearing involved complete removal of old palms. Late thinning allowed young palms to be underplanted, with no thinning until 8 years later when all old palms were removed. Gradual thinning involved underplanting and removing old palms over an eight year period. At the beginning, 12% of the old trees close to the young palms were removed; thereafter, the percentages removed by the end of each year were: 6%, 10%, 12%, 18%, 10%, 12% and 8% for years 1, 2, 3, 4, 5, 6, 7, and 8, respectively.

Late thinning resulted in retarded growth and development of young palms; only 44% of young palms had flowered and 83% were non-bearing at the time the old trees were removed. However, late thinning

Figure 2.10. Young palms interplanted between
the rows of older palms; Sri Lanka.

Figure 2.11. Two and one-half year old palms underplanted in older palms, with newly established _Brachiaria brizantha_ pasture planted between the palms. Western Samoan Trust Estates Corporation.

Figure 2.12. Young palms underplanted in old, hurricane-damaged grove. Mulifanua Plantation, Western Samoan Trust Estates Corporation.

resulted in the highest total nut yield for the 12 year period studied, indicating the significant contribution to yield by the old palms.

From the standpoint of pasture improvement and use, gradual thinning would appear to have some advantages over late thinning. Gradual thinning would allow redesign of the planting layout, making it possible to use group plantings or strip plantings of new palms by selective removal of old trees. Pieris (1945) and Lambert (1970) gave detailed instructions on methods of thinning to obtain desired numbers of trees per ha. These instructions could be modified to allow thinning to obtain desired planting zones in old stands and to accommodate planting designs most suitable for coconut/pasture/cattle enterprises.

For cost estimates of underplanting in Sri Lanka see Cheyne (1952), and for a discussion of management problems relating to underplanting see Rodrigo et al. (1952).

REHABILITATION OF EXISTING GROVES

One advantage of coconut/pasture/cattle farming is the opportunity to obtain additional income to rehabilitate rundown or damaged plantations. With severe injury or even death of palms caused by lightning, hurricane, disease or other calamity, thin stands and even large clearings 0.5 ha or more in size may result. Such gaps in stands should be replanted, underplanted, or used for intercropping.

Many small farmers allow their groves to become overcrowded (Figure 2.9). Sometimes this is the result of planting too many trees at establishment of the coconuts; more often uncollected nuts sprout on the ground, take root, and in the absence of any thinning steps, grow to become mature palms. Some farmers mistakenly believe that the more palms, the higher the yield. Unfortunately, this is not true. Palms in such groves are randomly spaced, and plant populations may reach as high as 275 to 350 palms per ha or more. This is far too many, and the stand should be thinned to about 125-150 palms per hectare to reduce competition between palms, to raise copra yields, and to allow good growth of intercrops or pasture.

In thinning, poorest trees should be cut first. The remaining palms should then be selected in regard to distance from nearest palms, vigor, age and general condition. Lambert (1970) has outlined a straight-forward approach to thinning using a system of stakes and strings to lay out desired rows and spacings (on a grid basis) in a grove. He then suggests selecting and marking the best trees on or near the desired rows and spacings and thinning the rest.

If thinning is done in areas infested with rhinoceros beetle, the palm trunks and other field debris must be gathered and burned, to reduce breeding sites. This may be no small task. For example, the Philippines estimates that if current steps to replace old and senescent palms are carried out, six million old palms per year will be cut down and destroyed or used for other purposes such as posts or special wood products (Philippine Coconut Authority, 1975).

3
Understory Cover Management in Coconuts

All coconut producers face the problem of deciding how to handle the land area under the palms. Because coconut is such a long-term crop, there are only a few options available, ranging between uncontrolled growth of natural vegetation under the palms, controlling vegetation to the extent that nuts can be easily seen, or even cultivating to keep the soil surface bare and weed-free. All options in the range cause problems, but the extremes cause great difficulty. Uncontrolled vegetation growth is unacceptable because the understory becomes a "jungle" of brush and rampant weeds, thereby denying the producer access to his crop and virtually assuring that nut collection will be incomplete, difficult, and expensive. On the other hand, repeated cultivation or clean weeding will result in near-perfect nut collections, but can also result in soil erosion and rapid runoff of rainfall. Cover management and control will be necessary; the question is; what kind of cover management is best for the system?

Several ways are used to control the cover under coconuts. These include: (1) intercropping with food or cash crops (discussed in Chapter 4); (2) controlling weeds by hand slashing (sometimes called hand brushing or cutlassing), or mowing, to keep understory growth short enough to reduce competition with the palms and to make nut collection easy; (3) combinations of chemical vegetation control (herbicides) and mechanical control; (4) periodic burning to clear away and reduce understory vegetation; (5) planting the understory to leguminous cover crops; and (6) grazing of cattle (Osborne, 1972) or other animals (see Chapters 5, 7, 9, and 10). Methods 2, 3, 4, and 5 will be discussed more fully in following sections of this chapter.

WEED CONTROL

Of all the problems faced by the coconut farmer, controlling weeds is the most difficult and expensive. Weed control is a never-ending process and must continue year in and year out over the life of the crop. For many farmers, grazing livestock under the palms to reduce weed competition is the cheapest and easiest way to overcome their weed control dilemna. Farmers must control weeds, otherwise the understory becomes choked with herbaceous and woody plants, and palms suffer greatly from competition for both nutrients and moisture.

Weed and brush encroachment and types of weeds

Natural weed infestations in coconuts include a wide variety of perennial grasses, forbs (herbaceous broad-leaved plants), ferns, shrubs, and even small trees (Figures 3.1, 3.2, 3.3, 3.4). Most serious weeds are hardy perennials that grow and compete in the open ground areas beneath the palms. Few annuals are important weeds in coconut, except during the early stages of development of the grove. No one plant predominates, probably because of the widely-varying areas throughout the tropics in which the crop is grown and because of the many versions of intercropping or companion enterprises under which the crop is grown. Weed control in coconut is difficult because it is mostly a matter of post-emergence control of established, hardy, and even rampant plants.

Holm, et al. (1977) in a study of the worst weeds of the world, compiled a list and rankings of importance (serious, principal, common, or "weed") of weeds in coconut (Table 3.1). Many of these weeds are pasture plants; in fact, many of them are important in natural or improved pastures under coconuts (see Chapters 5, 7, and 8). Their presence in such conditions indicates a certain amount of shade tolerance.

Grasses which present problems are: Imperata cylindrica (Anonymous, 1958), Axonopus compressus, Paspalum conjugatum, Panicum repens, Panicum maximum, and Brachiaria mutica.

Herbaceous weeds (forbs) which are problems are
Mimosa *pudica*, *Mikania* *cordata*, *Passiflora* *foetida*,
Bidens *pilosa*, *Synedrella* *nodiflora*, *Sida* *acuta*, and
Solanum *torvum*.

Shrubs or small trees become a problem in either
young or older plantations that are poorly managed
or carelessly grazed. Important weeds are: *Chromo-*
laena *odorata* (syn. - *Eupatorium* *odoratum*), *Psidium*
guajava, *Lantana* *camara*, *Stachytarpheta* spp., and
Mimosa *invisa*.

Discussion of Individual Weeds

Some of the weeds in Table 3.1 are discussed
in detail in Chapter 5 (*Axonopus* *compressus*, *Cynodon*
dactylon, *Imperata* *cylindrica*, *Mimosa* *pudica*,
Panicum *repens*, *Paspalum* *conjugatum*, and *Ruellia*
prostrata) and Chapter 8 (*Brachiaria* *mutica* and
Panicum *maximum*). The remainder of the weeds in
Table 3.1 will be discussed in the following section of this chapter. Common names used are those
suggested by the Weed Science Society of America
(1966).

Figure 3.1. Weed growth under older palms.
Note large number of nuts on the ground.
Weeds here are grasses, forbs and small
shrubs which are mowed or slashed down
periodically.

48

Figure 3.2. D. E. F. Ferdinandez standing in weed growth in a palm nursery at the Coconut Research Institute of Sri Lanka. Weeds here are shade-tolerant grasses and forbs.

Figure 3.3. Very dense undergrowth of weedy grasses, clambering vines, and small shrubs in 20 year old palms. Nut collection in such dense growth would be very difficult.

Figure 3.4. Natural pastures, badly infested with Nephrolepis ferns, WSTEC plantations, Western Samoa. These paddocks are burned every 3 years or so to control the ferns and to allow better pasture growth and production.

Figure 3.5. Workers hand-cutting Nephrolepis fern weeds with machetes; weeds are slashed down to a level just above the fallen nuts. Western Samoan Trust Estates Corporation.

Table 3.1. A list of some important weeds of coconuts and some of the countries
from which they have been reported (after Holm, Plucknett, Pancho and
Herberger, 1977; Holm, Pancho, Herberger and Plucknett, 1979; Pancho,
Vega and Plucknett, 1969; Nair and Chami, 1963; Lambert, 1973; Whistler,
Pers. Comm. 1974; Haigh, et al. 1951.)

| Weed | Ranking as to seriousness | | | |
	Serious (rank in top 3)	Principal (rank - top 5 or 6)	Common (top ten)	Weed (unranked or above rank of 10)
Ageratum conyzoides			Philippines Trinidad Malaysia	
Asclepias curassavica			Trinidad Western Samoa New Hebrides	
Axonopus compressus	Malaysia		Western Samoa	
Bidens pilosa		Trinidad		
Brachiaria mutica		Trinidad		
Cassia tora		Western Samoa	Fiji New Hebrides	

Table 3.1. continued

Weed	Serious (rank in top 3)	Principal (rank - top 5 or 6)	Common (top ten)	Weed (unranked or above rank of 10)
Chromolaena odorata (formerly Eupatorium odoratum)	Sri Lanka	Trinidad		
Clerodendron fragrans		Western Samoa		
Cynodon dactylon			Trinidad	
Cyperus melanospermus (also called C. aromaticus).		Solomon Islands Western Samoa		
Elephantopus tomentosus			Fiji Philippines	
Euphorbia hirta			Trinidad Western Samoa	
Imperata cylindrica	Sri Lanka	Papua New Guinea	Philippines	East Africa
Lantana camara		Western Samoa Fiji New Hebrides	Philippines	

Weed	Serious (rank in top 3)	Principal (rank - top 5 or 6)	Common (top ten)	Weed (unranked or above rank of 10)
Mikania cordata, M. micrantha			New Hebrides Western Samoa Fiji Solomon Islands Western Samoa	Philippines
Mimosa invisa		Papua New Guinea Western Samoa	Philippines	Sri Lanka
Mimosa pudica		Papua New Guinea Trinidad	Philippines Solomon Islands New Hebrides	Sri Lanka
Momordia charantia			Trinidad New Hebrides	
Nephrolepis exaltata; N. spp, including N. biserrata and N. hirsutula		Western Samoa Fiji	Solomon Islands	

Table 3.1, continued

Weed	Serious (rank in top 3)	Principal (rank – top 5 or 6)	Common (top ten)	Weed (unranked or above rank of 10)
Panicum maximum			Trinidad	Venezuela
Panicum repens	Malaysia			Sri Lanka
Paspalum conjugatum	Malaysia		Solomon Islands Western Samoa	
Passiflora foetida			Papua New Guinea	Philippines Trinidad
Pseudelephantopus spicatus				Western Samoa Philippines
Psidium guajava		Fiji		Trinidad New Hebrides
Ruellia prostrata				Western Samoa
Sida acuta			Western Samoa	Trinidad Solomon Islands

54

Weed	Serious (rank in top 3)	Principal (rank - top 5 or 6)	Common (top ten)	Weeds (unranked or above rank of 10)
Sida acuta				New Hebrides Philippines
Sida rhombifolia				Philippines Solomon Islands
Solanum torvum	New Hebrides	Fiji	Malaysia Solomon Islands	
Stachytarpheta cayennensis			Trinidad	
Stachytarpheta jamaicensis			Trinidad Philippines Solomon Islands Western Samoa	
Stachytarpheta urticaefolia			Fiji	

Table 3.1, continued

Weed	Serious (rank in top 3)	Principal (rank - top 5 or 6)	Common (top ten)	Weeds (unranked or above rank of 10)
Synedrella nodiflora		Trinidad Western Samoa	Philippines	
Urena lobata			Fiji	

<u>Ageratum</u> <u>conyzoides</u> and <u>A</u>. <u>houstonianum</u> (Tropic
ageratum, billy goat weed)

These two weeds may be easily confused. They
are hairy annual herbaceous weeds with terminal or
axillary flower heads. <u>A</u>. <u>houstonianum</u> has larger
flower heads and blue flowers; <u>A</u>. <u>conyzoides</u> (Figure 3.6) has faint blue, violet, or white flowers.
Both reproduce from seed and are troublesome in
plantation crops and pastures. They are quite shade
tolerant and colonize rapidly in disturbed or cultivated land.

Figure 3.6. An <u>Ageratum</u> <u>conyzoides</u> plant. <u>A</u>.
<u>houstonianum</u> closely resembles <u>A</u>. <u>conyzoides</u>
but differs in size and color of flowers.

57

<u>Asclepias</u> <u>curassavica</u> (bloodflower milkweed, red flower milkweed).

An erect, small perennial shrub 0.5 to 1.3 m high (Figure 3.7), <u>A</u>. <u>curassavica</u> is common in poorly managed pastures and waste areas. It can be identified by its purplish stems, bright red or orange flowers born in flat-topped clusters, rather long and lance-shaped leaves, and the long cylindrical fruit. It reproduces from seed. It is considered as poisonous to livestock, but usually is not palatable and so it increases when pastures are short of forage and are overgrazed.

Figure 3.7. <u>Asclepias</u> <u>curassavica</u>: plant habit; lower left and center - two views of the flower and side view of the tufted seeds.

Bidens pilosa (Beggar-ticks, Spanish needle)

Beggar-ticks is an erect annual herbaceous weed (Figure 3.8) which spreads readily from seed. The plant gains its common and generic names from the long slender seeds which have two hooked barbs at one end; these barbs catch on fur, cloth or other surfaces and thus the seeds are spread. The plant is easily recognized by its terminal yellow flowers and by its clusters of seeds which radiate in all directions from a common receptacle, giving a round, burr-like appearance to the seed heads.

Figure 3.8. Bidens pilosa plant with characteristic flower clusters and burr-like seed heads. At lower right is a close-up view of a seed with twin barbs at the tip.

<u>Cassia</u> <u>tora</u> (Foetid cassia, stinking cassia)

     An erect, branched herbaceous to somewhat woody small shrub up to 150 cm tall, <u>Cassia tora</u> (Figure 3.9) reproduces by seed. It is an annual or short-lived perennial, and can be readily recognized by its bright yellow flowers, sickle-shaped pods and compound leaves which, when crushed give off a foul odor. It is a serious weed of pastures, is unpalatable to livestock, and is difficult to control because of a deep tap root.

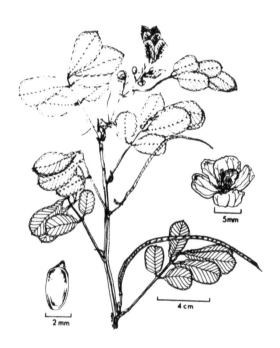

Figure 3.9. <u>Cassia</u> <u>tora</u>. Note curved, sickle-shaped pods and pinnately compound, round-tipped leaves.

Chromolaena odorata (Chromolaena)

Chromolaena odorata (Figure 3.10) formerly called Eupatorium odoratum, is a vigorous scrambling shrub which is a serious weed of plantation crops and pastures, especially in Asia. It cannot survive in heavy shade, but can withstand partial shading or full sunlight. It quickly invades open areas and forms dense thickets.

Chromolaena has yellowish stems, dark green opposite leaves with three conspicuous veins and toothed margins, and flat-topped clusters of light blue, light purple or white flowers. The leaves give off a pungent odor when crushed.

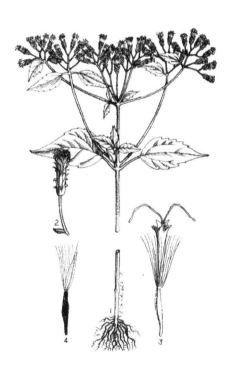

Figure 3.10. Chromolaena odorata; (1) partial plant habit, (2) flower head, (3) flower, (4) seed. (Illustration courtesy Professor Juan V. Pancho, University of the Philippines at Los Banos.)

<u>Clerodendron</u> <u>fragrans</u> (clerodendron, Honolulu rose)

A medium-sized, branched, spreading shrub with large ovate leaves and large terminal hydrangea-like clusters of showy white or pale pink flowers, <u>Clerodendron</u> (Figure 3.11) has excaped from ornamental cultivation to become a troublesome pasture weed. A native of southern China, it naturalizes easily and forms large thickets or patches on roadsides, and in waste places and pastures in the wet tropics. It is a declared noxious weed in Western Samoa and a problem weed in Fiji.

Figure 3.11. <u>Clerodendron</u> <u>fragrans</u> in a pasture in Western Samoa: The showy, rose-like flowers are characteristic of this weed.

Cyperus aromaticus (also known as C. melanospermus).
Navua sedge.

An erect, smooth perennial sedge, C. aromaticus
has become a serious weed of pastures in Fiji and
the Solomon Islands. It spreads rapidly and can
smother even vigorous forage plants such as para
grass. The plant is usually 30 to 60 cm high but
may reach 1.5 to 1.8 m in height. It can be dis-
tinguished by the cone- or button-like cluster of
flowers at the apex of a 3-angled flower stalk, and
subtended by 6 leaf-like bracts (Figure 3.12), three
of which are long (7.5 to 15 cm) and three are short
(2.5 to 6 cm).

This sedge is very difficult to control and is
unpalatable to stock.

Figure 3.12. Cyperus aromaticus in a pasture
in Western Samoa. Note round terminal inflorescence with numerous long leaf-like bracts
radiating from the base. This weed is very
competitive and is difficult to control.

Cyperus rotundus (purple nutsedge, nutgrass)

A small, grass-like, smooth, perennial sedge
(Figure 3.13), C. rotundus has been ranked as the
world's worst weed (Holm et al. 1977). It reproduces from seed and underground tubers and runners
which multiply rapidly and can survive for long periods in the soil. Somewhat shade tolerant, it increases rapidly in areas where herbicides are used
repeatedly, for it can survive topkilling and spread
into any existing void. It can be identified by its
triangular stem, the purplish flower heads borne at
the top of the stems, and by the underground tubers
and runners.

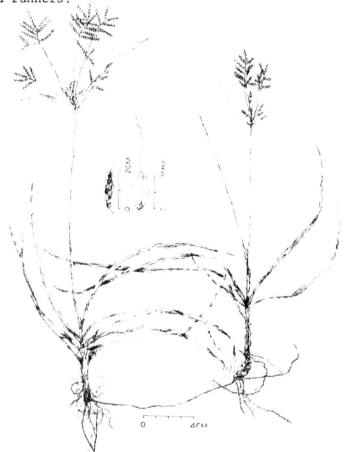

Figure 3.13. Cyperus rotundus, plant habit,
spikelet and seed.

<u>Elephantopus</u> <u>tomentosus</u> (elephantopus, elephant's foot)

Also known as <u>E</u>. <u>mollis</u>, elephantopus (Figure 3.14) is a serious weed of pastures and crops in the humid tropics. An erect annual which can be somewhat woody, it has hairy stems which arise from a rosette. Lower leaves are large, hairy and dark green in color. Flowers are small, white, borne in loose terminal clusters on much-branched stalks. The seeds are slightly barbed and can become attached to clothing or coats of animals. Wind may also disperse the seeds. Elephantopus spreads rapidly in cultivated lands and in pastures; it is not palatable for animals.

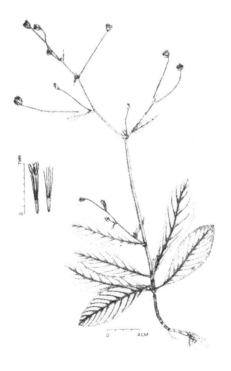

Figure 3.14. <u>Elephantopus</u> <u>tomentosus</u>; plant habit; left center - flower, seed.

Euphorbia hirta (Garden spurge)

Garden spurge is a small, prostrate annual weed (Figure 3.15) with reddish hairy stems, leaves with notched edges, and milky sap. Somewhat shade tolerant, it can withstand mowing or cutting but is useless for livestock feed.

Figure 3.15. An Euphorbia hirta plant.

Lantana camara (Lantana)

Lantana is a spreading, thicket-forming woody shrub (Figure 3.16) which can be troublesome in pastures and perennial tropical crops. It is some-what shade tolerant and will grow profusely under coconut. The seeds are carried by birds which can spread it widely. Lantana can be readily distin-guished by its flat-topped clusters of flowers which are generally yellow and pink on opening but change to orange and red, or sometimes blue or purple. The nearly round fruits are dark purple or black in color. Lantana can cause photosensitization or gastrointestinal disorders in grazing animals.

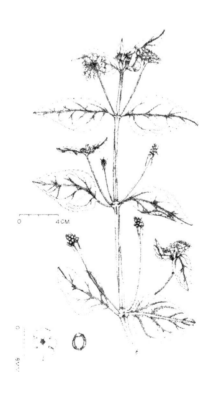

Figure 3.16. Tip of a Lantana camara branch. Flower clusters are shown at the top and lower right; fruits are shown at the center and lower left.

Mikania cordata and M. micrantha (Mikania, mile-a-minute)

Mikania is a rapid-growing, creeping or twining, rampant perennial vine (Figure 3.17) which smothers young trees and other plants. There is some confusion about the distribution and correct names for these weeds in Asia, Oceania, and Africa (Holm, et al, 1977). It can tolerate light to moderate shade but not heavy shade.

A component of some lightly grazed natural pastures, it cannot stand close continuous grazing.

The vines have cordate leaves with 3 to 7 veins originating from the leaf base and white flowers borne in nearly flat-topped clusters. It reproduces by seeds with feathery bristles which act as a parachute, carrying the seeds for long distances in the wind.

Mikania can be grazed or cut and fed green to animals. It was spread widely in the world by plantation owners who used it as a cover crop.

Figure 3.17. Mikania cordata - closeup photograph showing characteristic somewhat heart-shaped or arrowhead-like leaves.

68

Mimosa invisa (Giant sensitive plant)

A shrubby, somewhat herbaceous, spreading, thicket-forming plant (Figure 3.18), giant sensitive plant becomes woody with age. It somewhat resembles and is sometimes confused with its relative sensitive plant (Mimosa pudica), which is much more sensitive to touch.

Distinguishing characteristics of the plant are: the bipinnate compound leaves; angular stems with recurved spines or thorns; round, fluffy, puffball flower heads which are light pink in color, and the 3- to 4-seeded pods which are borne in clusters.

The plant invades coconut lands and, being unpalatable to animals because of the spiny stems, it often becomes dominant, forms dense thickets, and is difficult to control. It is a very troublesome weed in pastures.

Figure 3.18. Mimosa invisa: (1) plant habit (2) seed, (3) cluster of pods, (4) pod, (5) leaflet, (6) flower. (Drawing courtesy of Professor Juan V. Pancho, University of the Philippines at Los Banos.)

69

<u>Momordica</u> <u>charantia</u> (balsam pear, bitter melon)

   An annual prostrate, creeping or climbing vine
(Figure 3.19), M. <u>charantia</u> is sometimes cultivated
for its rather bitter fruits which are used in Asian
dishes. It can be easily identified by its growth
as a vine, deeply lobed leaves, bright yellow flowers and 6 small (10-12 cm long) rough or warty
fruits which are grooved longitudinally. It repro-
uces from seed.

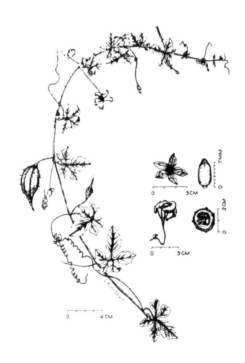

Figure 3.19. <u>Momordica</u> <u>charantia</u>: plant
habit, flower, cross section of fruit, seed.

70

<u>Nephrolepis</u> <u>exaltata</u>, <u>N</u>. <u>biserrata</u>, <u>N</u>. <u>hirsutula</u>
(Nephrolepis, sword fern.)

Nephrolepis is an upright creeping fern (Figure
3.20) with long wiry rhizomes. The fronds are upright, stiff, light green, 0.3 to 0.6 m long, about
10 to 12 cm wide, bearing numerous lance-shaped
leaflets at right angles from the axis, and somewhat
hairy on the midrib and undersurface. The sori
(spore-bearing bodies) are borne near the margins of
the undersides of the leaflets.

This fern and its relative, <u>N</u>. <u>biserrata</u>, are
problem weeds in many coconut areas. It is difficult to control by herbicides, and many coconut farmers have resorted to the use of fire to control it.
It can be smothered and controlled by vigorous
creeping pasture grasses such as <u>Brachiaria</u> <u>brizan-</u>
<u>tha</u>.

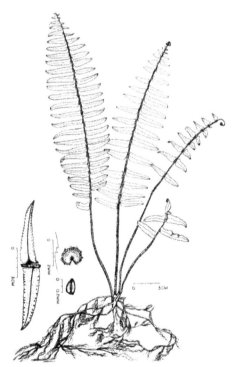

Figure 3.20. <u>Nephrolepis</u> <u>exaltata</u>; plant
habit, underside of leaves showing spore-
bearing sori, closeup of sori, and spores.

71

Passiflora foetida (wild passion fruit)

A perennial creeping vine (Figure 3.21), which climbs by means of tendrils. The leaves are alternate, hairy or bristly, more or less 3-lobed, and give off an unpleasant smell when crushed. The flowers are attractive -- white, purple or blue. The fruit is round, fairly hard, yellow, and bears many seeds. This weed is often found in drier, disturbed areas.

Figure 3.21. Passiflora foetida: plant habit, upper left - flower; lower left - fruit.

Pseudelephantopus spicatus (false elephantopus)

As its name indicates, P. spicatus is often mistaken for its relative, Elephantopus tomentosus. P. spicatus is a coarse, erect perennial herb (Figure 3.22) which may be somewhat woody. It has hairy stems which arise from a loose rosette of leaves which are usually much larger than those in upper parts of the plant. The leaves are dark olive green in color and are hairy. The flowers are small, white, borne in slender terminal spikes. The fruit is small, 6 mm or so long, slightly hairy, and slightly barbed. Seed production is very heavy, and the plant spreads rapidly in poorly managed pastures. It is unpalatable for livestock.

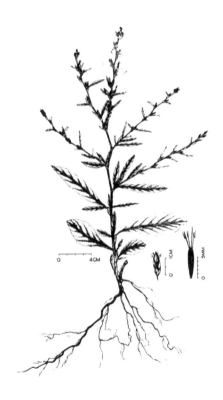

Figure 3.22. Pseudelephantopus spicatus: plant habit; flower, seed with small barbs at the tip.

Psidium guajava (guava, yellow guava)

A vigorous, rapid growing shrub or small tree
(Figure 3.23) generally under 8 m tall, guava is
cultivated in the tropics for its fruits which make
flavorful juices, jams and jellies.   In many places
it has escaped to become a serious weed in pastures
(Figure 3.23).   The tree has smooth, copper-colored
bark which sometimes peels, ovate leaves with very
prominent veins, and cream-colored or white flowers.
Fruits are oval or egg-shaped; yellow; 5 to 10 cm
long; the surface somewhat rough; the flesh is yellow or pink, bearing many seeds.

Guava spreads rapidly from its numerous seeds
which are disseminated by birds.   Dense stands can
rapidly cover a pasture and render it practically
useless.

Figure 3.23.   Psidium guajava; portion of
branch showing leaves and fruit; lower left -
longitudinal cross section of fruit; center -
seed; lower right -tip of branch showing
flower and bud.

<u>Sida</u> <u>acuta</u> (Southern sida, broom weed)

A small perennial shrub, near-shrub or herb (Figure 3.24) with tough woody stems and a deep tap-root, southern sida is widespread throughout the tropics and subtropics. Once established, the plants are very tough and competitive and can with-stand drought, trampling, mowing or shallow infre-quent tillage. In overgrazed pastures the plant increases rapidly. It reproduces by seed each of which has two sturdy awns at the tip; these awns stick to animals and spread the seeds and capsules. The flowers are light yellow, yellow or pale orange, with five petals; the fruit is a capsule bearing 5 to 8 seed segments which split at maturity. The leaves of the plant are small and lance shaped.

Figure 3.24. Sida <u>acuta</u>: 1) plant habit, 2) leaf, 3) flower, 4) fruit (capsule), 5) seed. (Drawing courtesy of Professor Juan V. Pancho, University of the Philip-pines at Los Banos.)

75

A close relative of S. acuta is Sida rhombifolia (Figure 3.25). It behaves much like S. acuta but can be differentiated by its wider rhomboid or diamond shaped leaves, and by its seed segments which have only one awn or more at the tip.

Both of these plants may be lightly grazed by animals, but are not very palatable.

Figure 3.25. Sida rhombifolia: plant habit, flower, capsule and seed.

Solanum torvum (prickly solanum)

An erect, robust perennial shrub 2 to 3 m tall, S. torvum is a dreaded weed in crops as well as pastures (Figure 3.26). It is best known for its spiny, somewhat hairy stems and spiny leaves. Leaves are large and more or less lobed; flowers are small, white with a cream center, and borne in groups of 5 and 6; and the fruits are smooth, yellowish, green, containing more than 250 small seeds. Seeds are spread by birds, men, and animals. It has been reported as a weed in many coconut-growing countries of the world, although not necessarily as a weed in coconut.

Figure 3.26. A Solanum torvum plant.

Stachytarpheta cayennensis (cayenne vervain), S.
jamaicensis, (Jamaica vervain), S. urticaefolia
(nettle-leaved vervain)

There are seven species of Stachytarpheta
which are classed as weeds. Of these, S. cayennensis, S. indica, S. jamaicensis (Figure 3.27) and
S. urticaefolia are probably most important
(Plucknett and Whistler, 1977). These are herbaceous or woody perennial shrubs which are problems
in tropical plantation crops. They may be differentiated mainly on the basis of plant size, flower
size and color, and leaf characteristics. They are
often confused with one another. Apparently the
species hybridize, creating even more diversity.

S. cayennensis has slender flower spikes, less
than 3 mm in diameter, and the fruits are sunk in
wider grooves, as wide as the spikes. The flowers
are blue or white.

S.urticaefolia (Figure 3.28) has deep blue,
rarely white flowers borne on spikes up to 50 cm
long.

All of the Stachytarphetas are unpalatable to
stock, reproduce profusely from seed, form dense
stands, and crowd out more desirable species.
Mowing is ineffective in control; therefore chemical control measures must be used.

Figure 3.27. Stachytarpheta jamaicensis:
left- plant habit; flowers, fruit

Figure 3.28. Stachytarpheta urticaefolia:
plant habit, flowers.

79

Synedrella nodiflora (synedrella)

An erect, branched, annual herb with woody stems (Figure 3.29), synedrella is a common weed in cultivated lands, including coconut.

About 30 to 70 cm tall, it has opposite leaves which are rough to the touch and with slightly notched margins, yellow flowers clustered in the leaf axils at the nodes of the stem, and a small seed with two barbs at the tip. It reproduces from seed. It is easily recognized because of the clusters of yellow flowers in leaf axils at the nodes.

Figure 3.29. Synedrella nodiflora: plant habit; lower, from left - flower head, flower, flower, seed.

Urena lobata (urena, hibiscus burr)

Urena lobata is grown in Africa as a commercial fiber plant. A perennial shrub growing 60 to 120 cm in height, it has white, hairy stems; 3 to 5 lobed leaves with somewhat toothed margins; pink, almost sessile solitary flowers borne in the leaf axils; and a characteristic burred fruit which is covered with rigid hooked bristles. The fruits have 5 segments, with one seed to each segment; the seeds are brown. Urena is a problem weed in pastures. The spiny burrs stick to clothing and coats of animals. The plant is not palatable for livestock.

Methods of Weed Control

Manual weeding. Probably the most common system practiced, in most cases manual weeding consists of persons equipped with long bush knives (called machetes, cutlasses, etc. in various places), who cut the standing vegetation down to a height of about 10-20 cm (Figure 3.5). This system may be repeated once or twice or even several times a year, depending upon growing conditions, labor availability, and diligence of the farmer.

Some farmers do pull some weeds by hand; these are usually troublesome forbs, shrubs, and small trees. Hand grubbing by mattock or hoe may be necessary for weeds with heavy root systems.

Regardless of the system used, manual weed control is onerous and expensive. When copra prices are low, groves are often neglected, leading to serious weed build-up, brush encroachment, and decline in palm condition and yield.

Mowing

For lands which are not too stony, steep, or otherwise difficult to move or operate machinery, mowing is a satisfactory measure in controlling weeds. It may be especially important in preventing seed production in troublesome weeds. Mowing in combination with chemical or manual control should be quite effective against brush and small trees. Mowing can also be a useful practice in management of grass/legume pastures to prevent grasses from becoming dominant.

Pastures as a weed control measure

Good pastures can reduce weed encroachment and weed control costs to a minimum, provided that good management is practiced. Goonasekera (1953a, 1953b) reported that, in a grazing trial under coconut in Sri Lanka, Sida acuta, Mimosa pudica, Vernonia javanica and Hedyotis auricularia invaded rapidly in heavily grazed (overgrazed) plots. Better pastures had fewer weeds.

Some improved tropical pasture legumes and grasses are very effective competitors with weeds. Notable among these are some of the rhizomatous or stoloniferous grasses such as Brachiaria brizantha, B. mutica, B. miliiformis, and Digitaria decumbens.

Chemical weed control

Most weedy grasses can be controlled by dalapon (2,2 dichloropropionic acid) or TCA (trichloroacetic acid). Dalapon does not harm young coconuts if it is applied as a directed spray around the plants. Other herbicides which can be used to control established grasses in coconuts include diuron (N'-3, 4-dichlorophenyl)-NN-dimethylurea) paraquat (1, 1-dimethyl-4, 4'bipyridilium) and linuron (N'-(3, 4-dichlorophenyl) -methoxy-N-methylurea) (Hoyle, 1969). Of these materials paraquat, diuron plus a surfactant, and dalapon may be most satisfactory.

Broadleaf weeds and shrubs must be controlled in coconuts. Shrubs especially may become important in pasture/coconut enterprises. Romney (1964) found that 2,4-D (2,4-dichlorophenoxyacetic acid); 2,4,5-T (2,4,5-trichlorophenoxyacetic acid); and MCPB (4-chloro-2-methylphenoxy-butyric acid) were toxic to young coconuts when sprayed directly on the young palms. Dalapon at 6 kg/ha in 380 liters of water sprayed on the seedlings damaged the young trees. Later work by Romney (1965, 1971) showed that 2,4-D amine, 2,4-D ester, MCPB, and silvex (2 (2,4,5-trichlorophenoxy) propionic acid) killed young palms when they were sprayed directly on the plants. Coconuts were not affected by soil applications of 3.5 kg/ha monuron (N-4-chlorophenyl) - N'N'-dimethylurea), 3.5 kg/ha diuron, or 8 kg/ha atrazine (2-chloro-4-ethylamino-6-isopropyla-mino -1,3,5-triazine). Hoyle (1969) reported no injury to coconuts with monthly directed sprays of 2, 4-D plus dalapon. For brush control in coconuts, it therefore seems possible to apply 2,4-D, 2,4,5-T, and silvex safely, provided that the sprays are not applied on the coconuts.

Table 3.2 presents herbicide recommendations for coconut in the South Pacific (Lambert, 1973). Other weed control references for coconut include Kasasian and Seeyave (1968), Solomon Islands Ministry of Agriculture and Lands (Undated), and Steel (1977).

Burning. Fire is used to control weed growth under palms in several parts of the world. The fires are started deliberately to burn off rank understory vegetation, often to control certain troublesome weeds for which no other control measures are successful.

In Western Samoa, the Western Samoa Trust Estates Corporation (WSTEC) burns paddocks infested with species of Nephrolepis ferns, usually N. exaltata and N. biserrata, (see Figure 3.4) and guava (Psidium guajava) every three years or so to reduce weed populations and competition (Morris Lee, Pers. Comm., 1972). In some cases these weeds may be shrubs or small trees, but often they are ferns or rank perennial grasses. WSTEC estimates that it costs US $7 per ha to control weeds in this way. Only paddocks more than 5 years old are burned.

In Papua New Guinea, pastures are burned early in the dry season, both to prevent fires later in the season as well as to allow grasses to produce new fresh growth for grazing (Bruce Jeffcott, Pers. Comm., 1972).

Other countries reporting use of fire to control weeds include the Solomon Islands (Ian Freeman, Pers. Comm., 1972), Tonga (Siaosi Moengangango, Pers. Comm., 1972), Sri Lanka (Rajapakse, 1950; Tempany, 1950; and Salgado, 1961a, 1961b) and Tanzania (Child and Groom, 1964).

Burning has been condemned in several quarters because it: (1) can cause damage to the palms; (2) allows build-up of fire-resistant grasses such as Imperata cylindrica (Tempany, 1950; Rajapakse, 1950) and shrubs such as Mimosa spp. (Ian Freeman, Pers. Comm., 1972); (3) if repeated frequently, leads to

Table 3.2. Suggested chemical control measures for coconut in the South Pacific (Lambert, 1973).

Chemical	Stage of Crop	Rate of use	Type of Weeds Controlled and Period of Control	When to Apply for Best Results and Type of Application	Remarks
Dalapon	Established plantations	11 to 17 kg/ha in 930 liters of water plus surfactant	Annual and perennial grasses	When grasses are young and growing vigorously	Do not spray young palms or use for ring-weeding
Diuron	Ringweeding of young palms	4.4 kg sprayed/ ha or 44 g per 100 sq. m. To spray an 2.4 m diameter circle = 2.1 gm per plant = 252 g per ha of coconuts if 120 coconuts/ha	Annual weeds and grasses for up to 4 months	Apply to bare ground as a directed spray. Avoid contacting foliage with spray or drift. Apply during the wet season or when rain can be expected.	Best results obtained if applied to bare ground. If weeds have germinated but are less than 15 cm in height add 1% surfactant at 1.9 to 3.8 l per knapsack. If weeds are more than 0.5

Table 3.2. (con't)

Chemical	Stage of Crop	Rate of use	Type of Weeds Controlled and Period of Control	When to Apply for Best Results and Type of Application	Remarks
Diuron					m to 1.0 m high either first cut weeds by hand then apply Diuron or add a contact herbicide to Diuron for knockdown and residual weed control
2,4-D Amine	Established plantations	1.4 to 4.2 liters per ha	Most broadleaved weeds	When the weeds are young and prior to seeding	Use the higher rate on mature weeds. Note: 2,4-D amine will damage legumes

Table 3.2. (con't)

Chemical	Stage of Crop	Rate of use	Type of Weeds Controlled and Period of Control	When to Apply for Best Results and Type of Application	Remarks
MSMA 1/	Ringweeding of young coco-nuts	0.94 to 1.4 l in 150 l water	Annual and peren-nial weeds and grasses	As required. Direct spray on weed foli-age in a radius of 1 to 1.3 m around base of young palms	Avoid spray drift on to young palms. This could cause scorch to fronds and could damage buds
Paraquat	Ringweeding of young palms	1.4 to 2.8 l/ha in 150 l water	Most annual and perennial weeds for knockdown only with no re-sidual effect.	Apply directed spray. Avoid con-tacting foliage of coconut and spray weeds and grasses to point of run-off	Take care to keep spray off green plant tissue of co-conut seeding

1/ monosodium methane-arsonate

sulfur deficiency; and (4) causes soil erosion and deterioration.

Fire damage to palms is a frequent criticism; however, advocates of the practice point out that timing, intensity, and duration of fires can be controlled to reduce risk of damage or loss. Damage has been reported from the Solomon Islands in young "Malayan Dwarfs" (Ian Freeman, Pers. Comm., 1972), from Tanzania where annual fires cause damage in current as well as future crops (Child and Groom, 1974), from Sri Lanka where uncontrolled bush fires cause considerable damage to palms (Salgado, 1961a), and Tonga (Siaosi Moengangango, Pers. Comm., 1972).

Burning to control understory growth probably should be practiced only as a last resort, especially to control certain problem weeds. In most cases, farmers who practice good grazing or pasture management under palms should find little cause to resort to burning. In fact, perhaps one of the strongest cases for managing pastures under coconuts can be made on the basis of the vastly improved weed control afforded by careful grazing, thereby eliminating or reducing fire hazards and the need for burning to control unwanted vegetation.

It should be mentioned that occasional controlled burning may be beneficial in shallow, rocky soils which cannot be tilled or cultivated.

COVER CROPS

Cover crops are grown for purposes of soil conservation and soil improvement. An herbaceous leguminous cover on the ground beneath the palms (Figure 3.30) reduces soil erosion, acts as a "smother crop" to check weed growth, fixes atmospheric nitrogen in root nodules which upon decomposition may yield nitrogen for coconut, assists in increasing soil water-holding capacity as well as percolation of water in the soil, and -- through leaf fall and decaying plant parts -- builds up soil organic matter, thereby improving soil texture and fertility.

Many plants have been used as cover plants in coconut. Most of those recommended for use are

Figure 3.30. Tropical kudzu; <u>top left</u>: close-up of the plant; <u>top right</u>: as a cover crop; <u>bottom left and right</u>: being grazed by tethered dairy cattle in Sri Lanka.

legumes; of these tropical kudzu (Pueraria phaseoloides), centro (Centrosema pubescens) and calopo (Calopogonium mucunoides) are used most frequently. Other legumes used or recommended include cowpea (Vigna unguiculata), Glycine wightii (formerly G. javanica), lotononis (Lotononis bainesii), sirato (Macroptilium atropurpureum), Macroptilium lathyroides, Vigna marina, and Canavalia seriacea (Lambert, 1970). Sproat (1968) discussed several legumes which have value in Micronesia. These include Desmanthus virgatus, Vigna marina (the most important legume cover plant for coconuts on atolls), Dolichos hosei (also called Vigna hosei), Sesbania speciosa, Mimosa pudica, Mucuna stizolobium, Crotolaria juncea and Desmodium canum.

Cover crops should not compete heavily with coconuts for nutrients and water. In order to reduce competition, cover crops should be mowed or grazed to remove excessive topgrowth. Some cover crops do become important in grazing systems under coconuts (Figure 3.30).

Cover crops will deplete soil moisture more than clean-weeded plots, especially during the first two years of growth, but soil moisture tends to rise thereafter under cover crops because of reduced runoff and greater infiltration into the soil (de Silva, 1951). Early competition for soil moisture is believed to cause observed drops in coconut yields after the planting of cover crops.

Creeping cover crops are probably best for soil conservation and weed control purposes. The most important ones -- tropical kudzu, centro, and calopo -- are creepers which are also shade or drought tolerant. Strengths of individual leguminous cover plants will be discussed in Chapter 8. For detailed discussions of cover cropping, see Bunting (1926); Holland (1926, 1927); Sampson (1928); Anonymous (1929); Joachim and Kandiah (1930); Telford and Childers (1947); M. de Silva (1951b, 1961); Salgado (1951a); Child (1964); Villeman (1964); Jordan and Opoku (1966); Hartley (1967); Pomier (1967); Purseglove (1968); Manciot (1968); Sproat (1968); Anonymous (1972); and Bourke (1975).

4
Intercropping with Coconuts

A very old mixed farming system, intercropping in coconuts is practiced by farmers in many tropical countries. Intercrops range from basic staple foods to cash and export crops. Wide spacing of palms, together with their long life and great heights, have caused farmers to find ways to obtain a second crop on the land beneath the palms (Ohler, 1969, 1972). While a detailed discussion of intercropping is outside the scope of this book, some consideration of the intercropping factors and systems is necessary for any utilization of coconut/pasture/ cattle systems. For detailed accounts of various types and systems of intercropping see Patel (1938), Pieris (1944), John (1952), Krishna Marar (1953, 1961), Seshradi and Sayud (1953); Menon and Pandalai (1958), Balasundaram and Aiyadurai (1963), Celino (1963, 1964), Sethi (1963), Anonymous (1964), Mangarat (1964), Rodrigo and Mangabat (1964), Sahasranaman (1964), Narayanan and Louis (1965), Santhirasegaram (1966e, 1967a), Owen Jones (1967), Chalmers (1968), Kee (1968), Kin (1968), Kotawala (1968), Leach (1968), Pedersen (1968), Smith (1968, 1971), Anonymous (1971b), Hampton (1972), Satyabalan (1972), Vernon (1972), Anonymous (1973), Kannan and Bhaskaran Nambiar (1973), Sahasranaman and Menon (1973), Nair et al. (1974), Nelliat et al. (1974), Creencia (1975), Cuevas (1975), Empig (1975), Iglesia (1975), Bourke (1976), Gallasch (1976), Leach et al. (1976), Nair and Varghese (1976), and Shepherd et al. (1977).

There has always been some controversy about the feasibility and desirability of intercropping in coconuts, and there are published recommendations both for and against it. In part, the mixed atti-

tudes seem to be caused by the very different
ecological situations and levels of management in
which coconuts are grown, as well as a reluctance on
the part of those interested mostly in coconut
production to accept the idea of subsidiary enterprises on coconut lands.

There are several advantages of intercropping
with coconut, including: (1) increased farm income;
(2) increased food production; (3) increased stability for coconut farms through diversification and
less dependence upon unstable market prices for
copra, coconut oil, coir and other products; (4)
closer care and attention in managing the intercrop
-- including tillage, weed control, use of fertilizers, irrigation, etc. -- often lead to improved
growth and yield of the coconut palms; (5) because
some cover management is essential in coconuts, the
cover might as well be income-producing crops rather
than weeds; (6) young palms do not produce economic
yields for at least six to seven years or more,
therefore, "catch crops" for food or sale may be
used to help pay for the establishment phase of the
grove; (7) intercropping can provide better labor
utilization and rural employment throughout the
year; and (8) where disease, (Figures 1.2 and 1.3,
Chapter 1) lightning, or other events have produced
thin, uneconomical coconut stands, intercropping
can produce needed income and better land use.

Disadvantages of intercropping include: (1)
competition by intercrops with coconuts for water
or plant nutrients; (2) intercrops may harbor or
attract pests or diseases harmful to coconuts; (3)
because two products are being raised on the land
area, an increased demand for fertilizer will result; (4) the growth habit of some intercrops may
cause difficulty in some coconut management operations such as application of fertilizers or harvesting; (5) if palms are shallow-rooted, tillage and
cultivation required for intercrops may cause root
damage to the main crop.

COMPETITION

Understory crops will not compete with mature
coconuts for sunlight and air, but competition for
water and nutrients could be substantial and must
be carefully considered. In newly-established
groves, tall intercrops could shade the young palms
unless care is taken to allow open spaces for the

91

small coconuts.

Competition for Water

Obviously, crop competition for water will occur most frequently in low rainfall areas, during dry seasons, or during periods of drought. In dry areas intercropping (including pastures) must be curtailed or practiced on a seasonal basis or even abandoned if water stress in palms is likely to result. However, it must be stressed that most crops will not produce any more water stress in coconuts than will be produced by weeds, and that cover management to reduce understory competition also will be necessary in dry areas.

Santhirasegaram (1967a) divided the coconut lands of Sri Lanka into three rainfall zones for intercropping: "wet," "intermediate" and "dry." He concluded that the Wet Zone (about 1900 mm or more annual rainfall, with good distribution) was suitable for intercropping year-round (Table 4.1). In the Intermediate Zone he concluded that short term intercrops could be grown during each of the two monsoon seasons. In the Dry Zone he recommended use of short term crops during the single monsoon. This information could be used as a tentative guide for other areas with similar annual rainfall and distribution.

There is a dearth of information about critical rainfall levels below which intercrops and coconuts would compete seriously for soil moisture. However, it is reasonable to assume that, where rainfall is well distributed and exceeds 1800 mm or so, competition for water between coconuts and intercrops will not likely occur.

Under conditions where rains are highly seasonal, where coconuts are grown on sandy soils or on soils with low moisture-holding capacity, competition for water could become serious and the management system adopted should be adjusted to reduce moisture stress for both crops. In India where intercropped coconuts are often irrigated, coconuts have benefited from the improved water supply (Gowda, 1971; Pillai, et al, 1964).

Nutrient Competition

Many coconut soils are deficient in the major

Table 4.1. Suggested intercropping patterns for
coconut lands in Sri Lanka, according
to rainfall (after Santhirasegaram,
(1967a).

Rainfall Zone	Dry Periods	Crops
Wet (1900 - 3800 mm)	None	Permanent crops. (pasture, fodder crops, coffee, cacao, plantain, pineapple, cassava, rice).
Intermediate (1500 - 2500 mm)	Feb. and June to Aug.	Short term crops (cowpeas, green gram, black gram, ground nut, sesame, rice, sorghum, cassava, maize) during S-W and N-E (ground nut, cassava, chillies) monsoons.
Dry (1250 - 1900 mm)	Feb. to August	Short term crops (ground nut, cassava, chillies) during N-E monsoon only.

nutrients (N, P, K) and in some minor nutrients. Therefore, nutrient competition is a real possibility in many intercropping systems. Competition for N and K could be especially serious (see Chapter 2 for information on nutrient requirements of coconuts). To reduce competition for nutrients, the individual nutrient needs of both the coconuts and the intercrop must be satisfied. Competition for N can be satisfied to some degree by growing nitrogen-fixing legumes. Grain legumes are especially beneficial because: (1) they compete with or replace weeds under the palms; (2) they fix atmospheric nitrogen; and (3) they are a food crop. Competition for potassium could be a problem when intercrops which have a high K requirement are grown. Coconuts have a high K requirement (Chapter 2) and many soils do not supply enough K for good growth and yield of palms.

In summary, provided that fertilization of both intercrop and coconuts is practiced in accordance with their individual needs, there should be little competition between the crops comprising the mixed farming system.

DOES INTERCROPPING HARM COCONUTS?

This question is most frequently raised when intercropping is suggested. What evidence do we have either for or against intercropping? Published reports indicate that, provided adequate care is given to both crops, coconut usually prospers in mixed farming systems. Improved coconut yields following intercropping have been reported in several countries (Anonymous 1966, 1964; Kotalawala, 1968; Kuttappan, 1971; Narayanan and Louis, 1965; Rodrigo and Mangabat, 1964; Sahasranaman, 1964; Sethi, 1963). Beneficial effects to palms are usually attributed to improved fertilization and plant nutrition, weed control, and other improved cultural practices provided for the intercrop. Apparently, the increased attention given to the intercrop compensates markedly for possible coconut yield losses which might result because of competition between the crops.

EFFECT OF COCONUTS ON INTERCROPS

While effects of intercrops on coconuts usually receive most attention, the effect of coconut on intercrops is also important. Aside from competi-

94

tion for nutrients and water, the main effect of coconut on intercrops will result from shading by the palms. The degree of shading by coconuts will depend on age and spacing of palms and soil fertility (Santhirasegarem, 1966b). Very young or old palms will not shade intercrops severely. Because of the heavy canopy produced from about the 5th to the 20th year of growth, low levels of light reach the understory for growth of intercrops. Nelliat et al. (1974) presented a graph showing the approximate shading of coconuts at different ages (Figure 4.1). Steel and Humphreys (1974) found that light transmission at noon through 30 to 50 year old palms spaced 10 to 12 m apart was 77 to 80 per cent of full sunlight.

The amount of sunlight which passes through the canopy is highly dependent upon the height of palms and the angle of the sun's rays (Figures 4.2, 4.3). In tall older groves, as the sun moves across the sky, sunlight reaches the ground at different times, since the shadowed areas change as the day passes. Less light reaches the understory during early morning and late afternoon.

SELECTING INTERCROPS

There are several factors to consider in selecting intercrops for coconuts; these include: (1) demand or market for the product; (2) soil type; (3) shade tolerance of the crop; (4) rainfall or irrigation facilities available; (5) spacing of palms; (6) age and height of palms; and (7) nutritional requirements of the intercrop (Narayanan and Louis, 1965). Table 4.2 presents a remarkable array of crops used in intercropping under coconuts. Table 4.3 presents yields of intercrops in the Philippines, both under and outside coconut groves.

HOW IMPORTANT IS INTERCROPPING?

Although exact figures for intercropping are difficult to obtain, statistics for some countries indicate its importance. In the Philippines, 68 per cent of 1,230 coconut farmers surveyed grew food or cash crops under coconuts (Nyberg, 1968). In Fiji, 14 per cent of the 72,000 ha of coconuts are intercropped (Satyabalan, 1972).

95

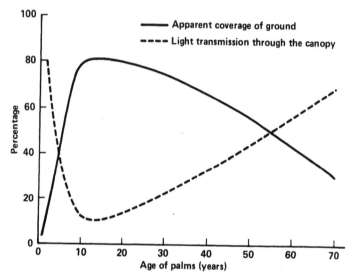

Figure 4.1 Apparent shading of ground by coconuts of different ages (after Nelliat et al. 1974).

Figure 4.2. Shading by 25 to 30 year old coconuts. Above - mid-morning; Below - mid-afternoon.

Figure 4.3. Shading patterns of palms: Above -
older palms (50-60 years) at mid-day; Below -
younger palms (about 30 years) in mid-afternoon.

98

Table 4.2. Some crops grown as intercrops in coconut (excluding cover crops, pasture, and fodder plants).

Crop	Scientific Name	Country (reference)
A. ANNUALS; SHORT OR INTERMEDIATE TERM		
1. Cereals		
a. ragi millet	Eleusine coracana	India (Albuquerque, 1964; Child, 1964) Sri Lanka (Child, 1964)
b. rice	Oryza sativa	India (Anon. 1971b; Child, 1964; Sahasranaman, 1964; Kuttapan, 1971)
c. maize	Zea mays	Philippines (Celino, 1963)
d. millets (unspecified)		India (Santhirasegaram, 1966e)
e. sorghum	Sorghum bicolor	Sri Lanka (Seshadri and Sayeed, 1953)

Table 4.2 (con't)

Crop	Scientific Name	Country (reference)
2. Grain legumes (pulses)		
a. mung bean	Phaseolus mungo	India (Child, 1964) Sri Lanka (Child, 1964; Seshadri and Sayeed, 1953, Philippines Anon. 1971).
b. green gram	Phaseolus aureus	India (Child, 1964)
c. cowpea	Vigna unguiculata	India (Child, 1964) Sri Lanka (Child, 1964)
d. ground nuts	Arachis hypogea	India (Albuquerque, 1964; Child, 1964; Santhirasegaram, 1966e)
3. Root crops		
a. cassava	Manihot esculenta	Sri Lanka (Child, 1964; Seshadri and Sayeed, 1953) Philippines (Celino, 1963) India (Anon. 1971b; Child, 1964)

100

Table 4.2 (con't)

Crop	Scientific Name	Country (reference)
b. sweet potato	Ipomoea batatas	India (Anon. 1971b, Child, 1964)
c. yams	Dioscorea spp.	India (Child, 1964; Gomez, 1974; Kuttapan, 1971)
d. taro	Colocasia esculenta	W. Samoa, Fiji, other Pacific Islands. India (Anon. 1964) Philippines (Gomez, 1974)
e. Elephant foot yam	Amorphophallus campanulatus	India (Anon. 1964)
f. tanier, cocoyam, yautia	Xanthosoma sagittifolium	Philippines
4. Vegetables (many crops)		Philippines (Celino, 1963) Malaysia India (Ghatge, 1964) Fiji (Hampton, 1972)

101

Table 4.2 (con't)

Crop	Scientific Name	Country (reference)
5. Condiments, Spices, specialty crops.		
a. Chillies	Capsicum annuum	India (Child, 1964) Sri Lanka
b. turmeric	Curcuma longa	India (Anon. 1964; Anon, 1971b; Balasundaram and Aiyadurai, 1963; Sahasranaman, 1964)
c. ginger	Zingiber officinale	
d. coriander	Coriandrum sativum	India (Anon, 1964; Balasun-daram and Aiyadurai, 1963)
e. fenugreek	Trigonella foenumgraecum	India (Anon, 1964; Balasun-daram and Aiyadurai, 1963)
f. sesame	Sesamum indicum	Sri Lanka (Seshadri and Sayeed, 1953)
g. black pepper	Piper nigrum	Philippines (Anon. 1971b)

Table 4.2 (con't)

Crop	Scientific Name	Country (reference)
6. Fibers		
a. cotton	Gossypium spp.	India (Albuquerque, 1964; Nyberg, 1968)
		Sri Lanka
B. TREES OR TREE-LIKE		
1. coffee	Coffea spp.	Malaysia
		Western Samoa
		Philippines (Celino, 1963)
2. cacao	Theobroma cacao	Papua New Guinea (Charles, 1959)
		Malaysia
		Western Samoa
		Fiji
		Philippines (Celino, 1963; Santhirasegaram, 1966e)
3. papaya	Carica payaya	India (Edachal, 1963)
		Philippines (Gomez, 1974)
		Western Samoa

103

Table 4.2 (con't)

Crop	Scientific Name	Country (reference)
4. banana	_Musa_ spp.	Philippines (Celino, 1963; Gomez, 1974) Malaysia India (Ghatge, 1964; Anon, 1966a; Kuttapan, 1971; Rodrigo and Mangabat, 1964; Sahasranaman, 1964) Caribbean (Child, 1964) Jamaica (Smith, 1968) Pacific Islands (Child, 1964)
5. plantain	_Musa_ spp.	Sri Lanka India (Nyberg, 1968) Caribbean (Child, 1964) Philippines (Child, 1964) Pacific Islands (Child, 1964)
6. abaca	_Musa textilis_	Philippines (Celino, 1963)
7. clove	_Eugenia caryophyllus_	Seychelles Islands Tanzania (Zanzibar)

Table 4.2 (con't)

Crop	Scientific Name	Country (reference)
8. cinnamon	Cinnamonum zeylanicum	Seychelles Islands Tanzania (Zanzibar)
9. arecanut	Areca catechu	India (Albuquerque, 1964; Paulose, 1965; Kuttapan, 1971)
10. mango	Mangifera indica	India (Albuquerque, 1964; Kuttapan, 1971)
C. OTHERS, NOT TREES OR TREE-LIKE		
1. pineapple	Ananas comosus	India (Anon, 1971b) Ceylon (Kutalawala, 1968; Seshadri and Sayeed, 1953) Kenya (Child, 1964) Philippines (Celino, 1963) Western Samoa
2. sugarcane	Saccharum spp.	India (Albuquerque, 1964; Nyberg, 1968) Philippines

Table 4.3. Yield of annual field crops under coco-
            nuts, Philippines   (after Paner, 1975).

Crop	Yield under coconut	
	Yield (T/ha)	Yield (%) com-pared to open
corn	1.17	37
soybean (Glycine max)	0.41	34
mungo (Phaseolus mungo)	0.58	85
peanut (Arachis hypogaea)	0.93	45
sunflower (Helianthus spp.)	0.63	95
jute (Corchorus spp.)	0.52	29
kenaf (Hibiscus spp.)	1.47	30
ramie (Boehmeria nivea)	0.16	135
taro (Colocasia esculenta)	5.48	367
cassava (Manihot esculenta)	0.98	14
arrowroot (Maranta arundinacea)	0.98	14

INTENSIVE INTERCROPPING SYSTEMS

Intercropping in coconut is not limited to just a single crop under the palms. In some cases several crops are grown together in complex systems. For example, in the Philippines pineapple and papaya are grown together in a 3-tiered system with coconut (Figures 4.4, 4.5). This system is one of the most profitable farming systems in the country; it can yield more than 100 per cent net profit over production costs (Gomez, 1974). In Western Samoa, cacao, papaya, and dadap (Erythrina lithosperma), a tree which is used as a green manure crop, are often grown together under coconuts. Figures 4.5, 4.6, 4.7, and 4.8 illustrate a variety of intercrops under palms.

Small farmers use intercrops and complex mixed cropping systems much more than larger land owners. Food crops, garden crops, fruits, and various cash crops are all grown together with coconut in the limited land area available. An idea of the number of crops can be obtained by examining an irrigated coconut farm in India described by Gowda (1971). The farmer owned a 12 ha irrigated farm in which coconut has been planted over the entire area. He grew sunn hemp (Crotalaria juncea) as a cover crop on half the land each year during the wet season and plowed it in. During the dry season, on this same land he grew horsegram (Dolichos uniflorus). The remaining 6 ha of land were used to grow sorghum, ragi millet (Eleusine coracana), and other subsidiary crops. He planted fruit trees such as sapote (Calocarpum sapota), guava (Psidium guajava), and cashew (Anacardium occidentale). He also planted border and windbreak trees such as Tectona grandis and Casuarina.

Figure 4.4. Intercropping under coconuts;
pineapple and papaya - Philippines.

108

Figure 4.5. Intercropping under coconuts.
Top -- pineapple, Sri Lanka;
Center -- pineapple, coffee, papaya, Philippines;
Botton -- pineapple and papaya, Philippines.

Figure 4.6.   Intercropping under coconuts:
top left -- sugarcane, Philippines; top right
-- sweet potato and plantain, Sri Lanka; bottom left -- plantain, Sri Lanka; bottom right
-- chillies and plantain, Sri Lanka.

Figure 4.7. Intercropping in coconut: top left -- cassava and plantain, Sri Lanka; top right -- foxtail millet, Sri Lanka; bottom left -- cacao, Alocasia macrorrhiza, and Erythrina lithosperma, a green manure crop, Western Samoa; bottom right -- robusta coffee, Western Samoa.

Figure 4.8. Intercropping under coconuts: top
-- bananas on north coast of Jamaica; bottom
-- cassava (tops lopped off by machete) with 6
year old palms, Sili Village Development Program, Savai'i, Western Samoa.

PLANTING AND SPACING NEW PALMS FOR INTERCROPPING
OR PASTURES.

For pasture production or intercropping wider
spacings between closely spaced groups of palms
could offer advantages. Grouped plantings would:
(1) allow normal numbers of palms per hectare to be
planted; (2) provide open unshaded areas for production of intercrops or pasture grasses and legumes; (3) lessen weed control problems in areas
directly beneath the closely spaced palms because
of more intense shading; (4) make fencing of young
palms less expensive because it would require only
a few groups to be fenced off rather than individual palms; and (5) provide essentially separate
land areas for the coconuts and the pastures,
thereby separating management practices for the
two crops. Under this system it should be possible
to use herbicides, to cultivate, or to slash down
weeds in the pasture areas with less effect on the
trees. Child (1964) tells of such a system, called
"bouquet" plantings, in Mozambique. The "bouquets"
are composed of 4 palms which are very closely
spaced, perhaps 3 to 4 or even 5.5 m apart in a
square pattern. Resulting open spaces between the
"bouquets" are then available for pasture. Of
course, variations on this system can be adopted.
For example, a 4.5 x 4.5 m square pattern would
give 177 palms/ha, and allow a 9 m space between
palm groups; and a 6 x 6 m triangular, strip-planted
system would provide 185 palms/ha and a 4.8 m
strip for intercropping.

There is great latitude for choice of spacings
and planting designs in order to make allowance for
pasture production under coconuts. Studies of
group or strip plantings should be initiated to
determine best combinations for pasture production.

"Hedge" planting for coconuts was proposed by
Liyanage (1955). He suggested over-planting by 30
per cent and then thinning within the rows by
roguing out weak or diseased palms. Under this
system rows could be spaced 8.5 m apart, and palms
could be planted 4.8 m apart in the row, alternating with trees in adjacent rows. Under this system it is necessary to remove old trees by the end
of the third year in order to reduce competition
for the young palms.

5
Natural Pastures

D. L. Plucknett and Flemming I. Ericksen

In most areas where grazing of cattle is practiced under coconuts, the pastures are composed of
"natural" species. Such species are weeds or adventive plants which spread without conscious effort by
man, but which must be controlled to reduce harmful
effects on coconut. Most "natural" pastures are
really weedy, untended lands which receive little or
no management. Animal production and coconut yields
are usually low.

In many countries cattle are used to graze down
understory weed growth and thereby to reduce weed
control maintenance costs. In the South Pacific,
cattle used in this way are called "brushers" or
"weeders". Some coconut producers raise cattle only
for weeding, and are reluctant to manage them for
production of meat or milk for fear that weed control benefits might be lost.

Since weed control is the single most important
management problem facing coconut farmers, use of
livestock to control weeds can be important. Osborne
(1972), citing costs for "manual brushing" versus
"cattle brushing" on a large commercial plantation
in the Solomon Islands, pointed out that manual
brushing costs U.S. $21 per ha per year and, assum-
ing that cattle took care of 75% of the weeding
needs, cattle brushing saved U.S. $14.80 per ha per
year in weeding costs. (For discussion of effects
of brushing on cattle production, see Chapter 9).

---

[1]Flemming Ericksen was formerly FAO Associate Expert
from Denmark; he worked for three years with the
UNDP Pasture/Livestock project in Western Samoa.
He obtained his Ph.D. degree in tropical agronomy
from the University of Hawaii. His address is:
Nylokke Vej 58, 8340, Malling, Denmark.

While some natural pastures are extremely weedy and low-yielding, there are some which can be quite productive (Figure 5.1). Natural pastures do benefit from improved pasture and livestock management; therefore, some discussion of the more common natural pasture species under coconut as well as management factors that affect carrying capacity and production will be presented.

EFFECT OF GRAZING ON PROBLEM WEEDS

With intensive grazing, such as is often practiced under coconut, certain weeds begin to decline in number and even disappear. Changes in weed flora result, leading to very different natural pastures. Some difficult or noxious weeds are controlled in this way.

Imperata cylindrica, for example, can be controlled by intensive grazing (Rajapakse, 1950; Anon.', 1958; Salgado, 1961a, 1961b; Charles, 1959). In Sri Lanka buffaloes have been used to control imperata, especially young plants, through close heavy grazing (Rajapakse, 1950; Anon., 1958; Salgado, 1961a, 1961b), including penning of animals in infested areas.

Other weeds which disappear with heavy grazing include Mikania cordata and most broad-leaved herbs. Plants which increase under heavy grazing include Axonopus compressus, Paspalum conjugatum, Mimosa pudica, Sida acuta, Cassia tora, Nephrolepis spp. (ferns), Desmodium heterophyllum, and Desmodium triflorum. Grazing intensity therefore may be used as a major management tool in regulating species composition of natural pastures.

NATURAL PASTURE PLANTS

Many weedy grasses, broadleaved weeds, shrubs, and even trees grow naturally under coconuts. A few of these plants have some forage value, but most are truly weedy -- of low palatability, competing with coconuts, and reducing forage quality and production. In some natural pastures weedy plants provide most of the poor quality forage available, and on which the animals are forced to subsist. When this occurs, the less palatable plants spread and become more dominant, and unless control measures are

115

Figure 5.1. A natural sensitiveplant (Mimosa pudica) pasture, Vailele Plantation, WSTEC, Western Samoa. Sensitiveplant is a favored plant for fattening cattle on WSTEC plantations, despite its reputation as a weed elsewhere.

taken, serious production losses for both crop and livestock will result. Some farmers believe that the grazing of cattle alone is sufficient to take care of their weed problems; this is not true and the belief should be vigorously discouraged. However, when coupled with selective weed control, grazing of natural pastures can be surprisingly productive and profitable.

IMPORTANT SPECIES

Many types of plants may be grazed and provide some feed for livestock. Grasses, broad-leaved herbs, sedges, and even shrubs are grazed; however, as would be expected, grasses and legumes are most important. A list of weeds of coconut was presented

116

in Chapter 2. Some of these are grazed. Also, cover crops such as tropical kudzu (Pueraria phaseoloides), centro (Centrosema pubescens) and calopo (Calopogonium mucunoides) can be found in natural pastures.

Important grass species include St. Augustine grass (Stenotaphrum secondatum), paragrass (Brachiaria mutica), carpetgrass (Axonopus compressus), bermudagrass (Cynodon dactylon), sour paspalum (Paspalum conjugatum), and guineagrass (Panicum maximum). Of these, perhaps carpetgrass and sour paspalum are found most widely and frequently. Imperata is found frequently in coconut, but is not of much value as a forage except when very young. Nevertheless, it is an important component of many natural pastures, especially where grazing pressures are low, where fires occur naturally, or where fire is used frequently by farmers to burn off rank, unwanted plant growth.

Important legumes are certain Desmodium species, sensitiveplant, Alysicarpus vaginalis, and brushy species such as leucaena (Leucaena leucocephala). Broadleaved herbs which are grazed include ruellia (Ruellia spp.), mikania (Mikania cordata), and commelina (Commelina spp.).

Axonopus compressus (Carpetgrass)

A creeping perennial, carpetgrass (Figure 5.2) is probably the most frequent and widespread natural pasture grass under coconuts. It can withstand very heavy grazing and trampling and is exceptionally shade tolerant. It is grazed in Fiji, Jamaica, New Hebrides, Sri Lanka, Solomon Islands, and Tonga. Carrying capacity of Axonopus pastures in the New Hebrides has been estimated at about 1.85 to 2.5 head per ha (R. Valin, Personal Comm.). Annual dry matter yield of a fertilized (150kg N, 40kg P, 80kg K per annum), predominantly carpetgrass pasture at Vailele, Western Samoa was 3360 kg/ha/year (R. Burgess, Personal Comm.).

Mimosa pudica (Sensitiveplant)

Often labelled a weed of coconut pastures, there has been much controversy about the value of sensitiveplant (Figure 5.3) as a pasture plant. That it becomes a dominant species after close grazing is unquestioned, but its value and even accept-

117

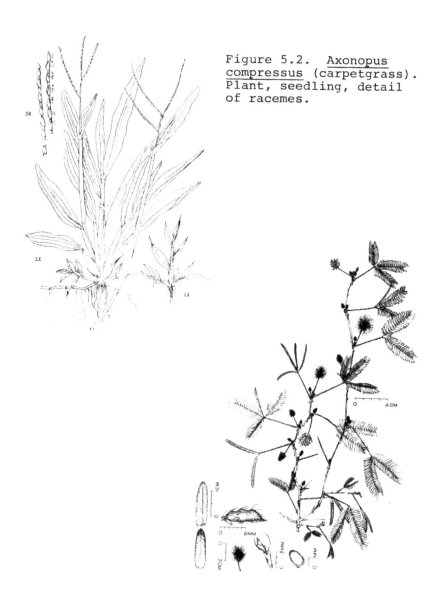

Figure 5.2. <u>Axonopus
compressus</u> (carpetgrass).
Plant, seedling, detail
of racemes.

Figure 5.3. <u>Mimosa pudica</u> (sensitiveplant);
plant habit; detail of seed, flower, leaflets
and pod. <u>M. pudica</u> is perhaps the most widespread legume in natural pastures under
coconuts.

118

ability as a forage plant have been debated. The Western Samoa Trust Estates Corporation (WSTEC) places great value on it, and has used predominantly sensitiveplant paddocks for fattening for more than 60 years (Morris Lee, Personal Comm.). WSTEC management considers that cattle fatten rapidly on the plant. In the Solomon Islands it is estimated that cattle obtain perhaps as much as half of their feed from sensitiveplant. Herman Retzlaff, an outstanding private coconut farmer in Western Samoa, has grazed carpetgrass/sensitiveplant pastures for years, averaging about 2.5 head to the ha. In his experience, cattle cannot graze it out, but goats can. However, there are indications in Samoa that under continuous, heavy, close grazing sensitiveplant may disappear.

Sensitiveplant has a tendency to cause clumping in pastures, especially in combination with grasses like paragrass. With heavier stocking and closer grazing and in combination with shorter grasses like sour paspalum or carpetgrass, it will not form clumps.

One objection to sensitiveplant is its spiny stems which make nut collection uncomfortable for workers with unprotected legs or feet. Also, some people believe that the spiny stems cause it to be unacceptable to animals. Another objection is that its spines can cause injury to the penis and pendulant prepuce of Bos indicus bulls. However, many farmers who have managed pastures in which sensitiveplant is dominant are enthusiastic about its acceptability and usefulness.

Sensitiveplant cannot withstand the dense shade of young coconuts. In ten year old palms near sea level in Western Samoa, ruellia and mikania were dominant under palms planted 9 X 9 m, while carpetgrass and sensitiveplant were dominant along roadways and in light shade. In nearby older palms (more than 30 years), where less shading occured, carpetgrass and sensitiveplant were dominant (personal observation, DLP).

Nut collection in carpetgrass/sensitiveplant pastures should be near perfect, for nuts can be seen easily at a distance on the low-growing, mat-forming grass/legume pasture (Figure 5.4). Sensitiveplant is adapted to a wide variety of soils, including somewhat poorly drained areas. It seeds profusely,

Figure 5.4.   Top:Carpetgrass (Axonopus
compressus)/sensitiveplant (Mimosa pudica)
pasture, Herman Retzlaff farm, Western Samoa.
Bottom:Carpetgrass/sensitiveplant/sour paspalum
(Paspalum conjugatum) pasture, Mulifanua
Plantation, WSTEC, Western Samoa.

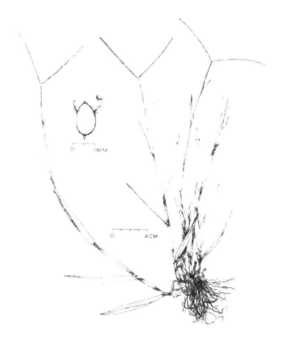

Figure 5.5. Paspalum conjugatum (sour paspalum)
plant habit and detail of flower. P. conjugatum
is probably the second most important grass in
pastures under coconuts.

establishes itself easily, and fixes nitrogen for
the pasture, thereby raising its own protein content
and contributing nitrogen to the soil.

Paspalum conjugatum (sour paspalum)

Perhaps second in importance as a natural
pasture grass under coconut, sour paspalum (Figure
5.5) becomes plentiful under heavy grazing pressure.
It apparently cannot stand as heavy grazing as
carpetgrass, however, and can be grazed out of the
sward.

Figure 5.6. Sour paspalum (<u>Paspalum</u> <u>conjugatum</u>)
pastures in Western Samoa. Top: With ruellia
(<u>Ruellia</u> <u>prostrata</u>), WSTEC. Center: In four
year-old palms. Bottom: With mikania (<u>Mikania</u>
<u>cordata</u>) in six year-old palms. Natural growth
such as this should be grazed to keep it under
control. Sili Village Development Program,
Savai'i.

Sour paspalum is a grass of the high rainfall tropics
(Figure 5.6). A creeping grass, its forage quality
is low. However, it does provide ground cover and
some grazing under low soil fertility. It is common
in closely-grazed pastures in most high rainfall
coconut lands in Western Samoa, Solomon Islands,
New Hebrides, Fiji, and Papua New Guinea (Charles,
1959).

Desmodium triflorum

This tiny legume (Figure 5.7) grows readily
under coconuts, but is too small to provide much
forage. Its contribution in natural pasture would,
if important, relate to its role in nitrogen fixation and in raising the protein level of the forage.
It is very common in coconut/pastures in Indonesia,
(Steel, 1974), Western Samoa and Malaysia (Verboom,
1968).

Desmodium heterophyllum (hetero)

Hetero (Figure 5.8) is found in many natural
pastures in the South Pacific. It is good ground
cover and has been used as a cover crop in some
tropical speciality crops such as pepper (Piper
nigrum). It has been recorded in lowland pastures
in Sri Lanka, Fiji, and Papua New Guinea. It is
strongly stoloniferous, establishes itself readily
in closely grazed swards, is adapted to a wide range
of soil conditions, is strain specific in its
Rhizobium inoculant requirement, and its nitrogen
fixation potential has been estimated at more than
130 kg/ha. It produces plentiful seeds in segmented
pods and is very persistent under close grazing.

Centrosema pubescens (centro)

Centro (Figure 5.9) has become naturalized in
many coconut areas, probably because of its usefulness as a cover crop in tropical tree crops. It is
a valuable pasture plant, and its presence in
natural pastures under coconuts will be a blessing
for the coconut farmer on whose land it grows.
Centro is shade tolerant. (For detailed information
on Centro see Cover Crops, Chapters 3 and 8).

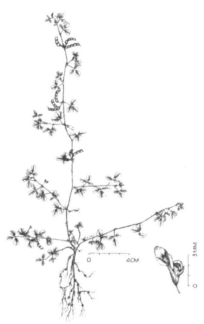

Figure 5.7.   Desmodium triflorum plant habit, flower.

Figure 5.8.   A Desmodium heterophyllum plant.
Note its very small circular leaves and
prostrate, creeping growth habit.

CENTROSEMA PUBESCENS BENTH.
Common Name - Centro

Figure 5.9. <u>Centrosema</u> <u>pubecens</u> (centro) plant, pods, seeds. (Courtesy Dr. Emil Javier, Plant Breeding Institute, University of the Philippines at Lao Banos).

Alysicarpus vaginalis (alyce clover)

A prostrate, mat-forming legume, alyce clover is hardy, deep-rooted, and drought-resistant. Characterized by its small, usually oval leaves, it spreads naturally by seed. It is common in coconut pastures in Sri Lanka (Haigh, et al., 1951).

Ruellia prostrata (ruellia, "blackweed")

This broad-leaved weed is grazed under palms in Western Samoa and provides forage in deep shade, especially under young palms. Whether ruellia is important elsewhere is not known, but its shade tolerance and apparent importance in Samoa make it worthy of mention.

Cynodon dactylon (bermuda grass)

Bermuda grass, a common creeping perennial, grows well at sea level in hot coastal regions. Although it is usually not very productive, it will survive under low rainfall in compacted soils.

Chrysopogon aciculatus

Chrysopogon (Figure 5.10) is not a very productive grass, but it is tough, is low growing and mat-forming, can be grazed closely, and resists trampling and abuse. Its barbed, prickly spikelets stick to clothing or other surfaces and can be annoying. It was formerly a favored pasture grass under coconuts in the New Hebrides (Weightman, 1977).

Stenotaphrum secondatum (St. Augustine grass, buffalo grass)

A creeping, mat-forming perennial, St. Augustine grass is tough and hardy. It resists close, heavy grazing, and is often associated with coastal pasture areas. It is a major grass in natural pastures in the New Hebrides (Weightman, 1977), where at least one farmer uses it to fatten stock at 30 months of age, often raising them in sour paspalum pastures.

MANAGING NATURAL PASTURES

Management systems for natural pastures mainly

126

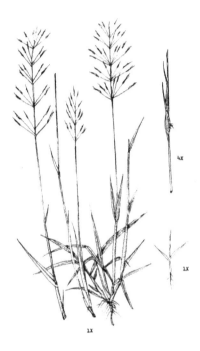

4X

1X

1X

Figure 5.10. Chrysopogon aciculatus; plant habit, seedling, detail of spikelet.

consist of grazing systems and livestock control
measures, supplementary weed control by mechanical
or chemical means, and -- in rare cases -- fertilization.

Grazing Systems

Specific grazing practices will be discussed in
Chapter 9; however, it should be pointed out here
that rotational grazing is well suited to coconut/
pasture systems. In part this is because the forage
can be grazed down low enough to allow unimpeded nut
collection, while at the same time allowing regrowth
of fairly good forage in rested paddocks. Weed control by grazing is considered by many to be much
better under rotational grazing than under continuous grazing, in which animals graze selectively,
thereby causing build-up of less palatable or desirable species.

Supplementary Weed Control

Since natural pastures usually consist of a
diverse mixture of plants, farmers must be alert to
prevent buildup of undesirable plants. Of particular concern are weedy shrubs and unpalatable perennial grasses that are difficult to control. Hand
cutting, grubbing, pulling or spot treatment with
suitable herbicides can all be used to control
unwanted plants.

Pasture Fertilization

Unless very desirable forage plants have become
naturalized in the pastures, in most cases it may
not be economical to fertilize natural pastures.
However, it is assumed that in most cases the coconut crop will be fertilized, whether or not the
pasture is fertilized. In cases where pasture yields
are so low as to cause concern, it is probably best
to consider planting improved pastures and to use
the fertilizer on the improved cultivated pastures,
rather than on unimproved natural pastures.

CARRYING CAPACITY OF NATURAL PASTURES

Osborne (1972) estimated that the carrying
capacity of unimproved pastures in the South Pacific
ranged from 2.4 ha/animal to 5 animals/ha. Of course
carrying capacities are highly dependent upon the

types of plants growing under the palms, rainfall, age of palms, degree of weediness, and other factors. In overgrazed, degraded lands carrying capacity is probably as low as one mature animal per ten ha or more. However, carrying capacities of 1.2 to 2.5 animals per ha are possible in many areas. Such figures illustrate the great potential for higher livestock production with improved pasture species and better management, even on natural pasture.

Table 5.1 presents some carrying capacities of natural pastures in a number of countries. For two of these countries, the number of animal grazing days per year was also recorded. It is impressive that most of the countries reported a carrying capacity of more than one head per ha for such pastures. Of course, livestock production figures were not presented for most of these countries, but such carrying capacities do indicate important potential of natural pastures under coconuts, as well as the potential for improved pastures.

Table 5.2 presents average carrying capacities of coconut farms in nine provinces on the island of Mindanao, Philippines (de Guzman, 1970). It should be pointed out that 76.6 per cent of the pastures under coconuts in this survey were natural pastures; the remainder was improved pasture. This, of course, would tend to raise somewhat the average carrying capacity values for each province.

Walton (1972) presented some very useful and helpful information on the effect of palm density and age on carrying capacity of natural pastures (Table 5.3). This effect is mostly related to the amount of shading caused by young, closely-spaced palms.

RESEARCH ON NATURAL PASTURES

Although several countries have initiated research programs on improved pastures few have made direct measurements on natural pastures. The pioneering research of the Coconut Research Institute of Sri Lanka evaluated coconut fields and herbage yields of the weedy natural growth under palms, but did not measure animal performance on natural pastures.

129

Table 5.1: Carrying capacity of natural pastures.

Country	Carrying capacity head(s)/ha	Number of palms per ha, (age)	Animal grazing days/ha/year	Reference
Jamaica	0.75	75 – 100	282	Coconut Industry Board, 1971
New Hebrides	1.5 – 2	very old palms, probably widely spaced	—	Weightman, 1977
Papua New Guinea (Morobe District)	2.5	—	—	Hill, 1969
Philippines Mindanao	0.5 – 1 1 animal unit/ 1.3 ha	—	—	Tanco, 1973
Solomon Islands	1.5 – 2 animal units (au) 1.25 au 1.0 au 1.25 au	175 215(4–14 yr.old) 260(4–11 yr.old) 260(11th yr.; thereafter)	— — — —	de Guzman, 1970 Walton, 1972 Freeman, 1972 Freeman, 1972
Sri Lanka	0.5 1.25 7–8 goats or sheep	— —	— —	Freeman, 1972 Ferdinandez, 1973 Ellewela, 1956
Trinidad and Tobago	0.75 – 2.5	125, average	—	de Guzman, 1974 Antoine, 1973
Western Samoa	2.5	125(20 yr.old)	427	Reynolds and Schleicher, 1975; Jensen, 1976
	1.5 – 2.0	125	expected annual liveweight gain= 220 kg/ha/yr.	Reynolds, 1977d

Table 5.2: Average stocking rate of 103 coconut/
beef cattle farms on the island of Mindanao,
Philippines (de Guzman, 1970).

Province	Stocking Rate
Agusan	1.2
Cotabato	2.3
Davao de Norte	1.6
Davao Oriental	1.9
Davao de Sur	1.1
Mindanao (average)	1.3
Misamis Occidental	0.7
Misamis Oriental	1.0
Zamboanga del Norte	1.0
Zamboanga del Sur	1.1

Table 5.3: The relationship of age of palm and
palm density on carrying capacity of natural
pastures in the Solomon Islands (After Walton,
1972).

Palm density, per ha	Age of palms, years	Carrying capacity, adult animal per ha
215	6 - 13	1.0
173	18 - 48	1.8
138	50 - 60	2.0

Jamaica

The Coconut Industry Board of Jamaica studied
the production of coconuts and cattle on both improved and native pasture. The species composition
of the natural pastures was unspecified. Natural
pastures were compared with pangola grass and guinea
grass. Results are summarized in Table 5.4.

Western Samoa

A major effort to evaluate the productivity of
natural pastures has been undertaken in Western
Samoa in a cooperative program between the UNDP,
Western Samoa Department of Agriculture and the University of the South Pacific School of Agriculture
at Alafua, Western Samoa. These trials have been
described in a number of reports (Reynolds, 1975;
1976a, 1978a; Reynolds and Ericksen, 1976; Reynolds
and Schleicher, 1975; Reynolds and Uati, 1976;
Reynolds, et al., 1978). The land was provided by
WSTEC and comprised part of Vailele Plantation.
The palms were 20 years of age when the grazing
trial began in 1972. Major natural pasture species
were: false elephantopus (Pseudelephantopus spicatus),
ruellia (Ruellia prostrata), sensitiveplant (Mimosa
pudica), Blechum pyramidatum, and Phyllanthus
urinaria. Lesser amounts of Ageratum conyzoides and
Mikania micrantha were also present. After six
months under fairly heavy grazing, false elephantopus,
Blechum, and mikania disappeared, while ruellia was
reduced by almost 50 per cent in one paddock but
less in another. Among plants that had increased
were sensitiveplant, carpetgrass, Cyperus kyllingia,
and Cyrtococcum trigonum.

During the first grazing period, August 1974 —
February 1975 (174 days), the natural pastures produced about 150 kg per ha liveweight gain, 365
animal grazing days per ha, and mean daily gain per
animal of 0.4 kg. On this basis the mean expected
gain per ha per year was slightly over 200 kg.
Forage yield and quality were higher than expected.
Dry matter yield for natural pastures was about
4300 kg per ha (guinea grass/centrocema pastures
averaged about 5200 kg per ha), and available forage,
both in regard to daily dry matter yield as well as
crude protein,were also quite high, 10.2 kg/animal/
day and 1.15 kg/animal/day,respectively. Reynolds
(1977d) summarized these year results from this trial
as follows (top of page 134):

132

Table 5.4: Pasture productivity under "Jamaica Tall" coconuts, 100 palms per hectare (Coconut Industry Board, Jamaica, 1971)

Pastures	Number grazing cycles (15 mo.)	Number cow grazing days/ha/yr.	Carrying capacity (cows/ha)	Number of nuts/palm (mean)
guinea grass	9.2	612	1.67	83.2
pangola grass	5.1	320	0.87	75.1
natural pasture	5.0	282	0.77	72.7
weed control	—	—	—	82.8

Trial period	Pasture	Stocking rate animals per ha	Liveweight gain/ha/yr
Aug.74–Feb.77	good "local"	2 - 2.5	181 [1]
Nov.76–Aug.77	moderate "local"	2 - 2.5	181 [2]

[1] palms 20 - 25 years old, light approximately 60% of open conditions

[2] palms 25 - 50 years old, light approximately 60 to 80% of open conditions

Solomon Islands

A grazing trial to compare improved pastures with native pastures was set up in the Russell Islands in 1973 - 74 (Watson, 1977). Main species present were sensitiveplant, centro, calopo and carpetgrass. Minor species included puero, hetero, Desmodium canum, mikania, sour paspalum, Synedrella modiflora, Borreria spp., and Nephrolepis biserrata. The pasture was fertilized with 20 kg P, 50 kg K, 20 kg S, and 0.15 kg molybdenum at the start, and 75 kg K was applied in November 1975. The pastures were continuously stocked at three stocking rates, 1.5, 2.5, and 3.5 animals/ha. Animal performance on natural and improved pastures is shown in Table 5.5. These results show that natural pastures do not have to be unproductive; indeed, if suitable species are present, production may be comparable with improved pastures.

Table 5.5: Animal performance on natural and improved pastures, Russell Islands, Solomon Islands (Watson, 1977; Whiteman, 1977).

Pasture	Stocking rate	Liveweight gains kg/head/year	kg/ha/yr
Natural	1.5	183	274
	2.5	161	402
	3.5	120	422
Improved[1]	1.5	179	268
	2.5	153	383
	3.5	128	447

[1] A mixture of Brachiaria mutica, B. decumbens, and B. humidicola, with centro, puero and Stylosanthes guianensis.

6
Planting and Establishing Pastures Under Coconuts

Establishing pastures under coconuts should be done in such a way that the best pasture can be developed with least injury to the coconuts. This will require planning as well as knowledge of the plants to be grown.

THE COCONUT ROOT SYSTEM

The coconut root system is composed of fibrous, branching roots which are attached to the base of the trunk or "bole." There is no taproot. Mature roots are stringy and slightly rope-like in appearance. Each tree has many roots, estimated at from 1500 to 2000 to as many as 4000 to 7000 for a mature palm. There are no root hairs. Absorption of water and nutrients takes place near the root tips, therefore, although palms have many roots, total absorptive surface is not great. Older brown-colored roots are strong and tough, lending support and protection for the trees. Under wet or water-logged conditions, special roots with whitish heads called "pneumatodes" are formed; these special roots are important in respiration (Ambrose, 1952). New roots are always being produced. It has been suggested that shallow tillage resulting in some root pruning will assist in increasing formation of new roots.

Roots grow both vertically and horizontally. Depth and breadth of root development depend largely upon soil and growing conditions. Horizontal feeder roots may extend 1.5 m or more away from the trunk. Diseased, poor-producing palms often have poorly-developed root systems (Michael, 1964).

135

LAND PREPARATION

Success or failure of planting pastures will depend heavily on the type and quality of land preparation methods used. Land preparation should be done as carefully and well as economics, equipment, labor supply, and soil conditions allow. For most planting systems, the existing vegetation must be removed or turned into the soil. To do this may require mechanical tillage, cutting or slashing by hand, use of fire, or combinations of these measures.

Manual Methods

For most small farmers, it will be necessary to clear and prepare the land by hand. In many countries fire is used to clear unwanted vegetation under the palms (see Chapter 3). Burning to clear the land may almost be mandatory for very stony soils where mechanical cultivation is impossible. Following burning it should be easy to prepare planting sites by hand.

The extent and quality of hand clearing and planting site preparation will vary also with the types and growth habits of pasture plants to be grown. Vigorous rhizomatous or stoloniferous grasses and legumes which spread naturally and rapidly to cover the ground will require much less land preparation than bunch grasses or upright legumes. Such plants can be planted at wide spacings; therefore they only require preparation of a few planting sites in the spaces beneath the palms.

Mechanical Cultivation and Tillage in Established Coconuts.

Disagreement exists as to the effect of tillage on coconuts, because some coconut roots are close to the soil surface and may be damaged by tillage implements. However, some producers and scientists believe some severing of roots is beneficial in stimulating new root growth. Many farmers do use disc harrows or plows of various kinds to kill weeds, to break up hardened soils, or to turn under crop residues or other vegetative materials (Figures 6.1, 6.2). In addition, there are many coconut lands which are too stony or rocky for satisfactory tillage. If perennial weeds with rhizomes are present, it may be necessary to cultivate several

Figure 6.1. Cultivation in coconuts; a
disc-harrowed field in Jamaica.

Figure 6.2. Disc-harrowing does cut some roots
of coconut. Most of the roots on the soil surface are coconut. The base of the palm is at
upper right.

times. It is difficult to determine how much or what type of tillage can be practiced in coconut.

Wijewardene (1954) recommended against cultivation in coconuts, suggesting that control of weeds or other covers should be done by rotary mowing. G. de Silva (1951) recommended careful plowing to break up "hardpans" and compacted soil; (1) to allow better aeration and infiltration of water; (2) to turn under weeds, green manures and other plant materials that improve soil texture, water-holding capacity, and fertility; and (3) to force deeper root penetration in order to reduce yield losses during drought. He recommended shallow tillage (15 to 20 cm deep) every 2 years, advising against plowing closer than 2 m from the base of a palm so that the fertilizer "manure circle" would not be disturbed. In Sri Lanka, plowing is usually done at the beginning of the rainy season and following application of fertilizers. To keep down weeds during the dry season, G. de Silva (1951) suggested shallow harrowing just at the end of the wet season.

Subsoiling of coconut plantations in Sri Lanka is sometimes practiced using heavy tines which work as deep as 0.6 m (Senanayake, 1952). It has been suggested that subsoiling should be done in alternate rows biannually, with subsoiler tines spaced 0.6 m apart (Wijewardene, 1954). This suggestion was not backed up by any evidence that the practice was needed. Table 6.1 shows data from Sri Lanka which suggest that subsoiling was beneficial for Paspalum commersonii pastures and associated coconuts, but that it had little effect on cori grass (Brachiaria miliiformis) pastures and coconuts.

A tractor and disc plow can be used to prepare seedbeds for: (1) sowing of grass and legume seeds, (2) spreading of vegetative cuttings of grasses (infrequently of legumes) and discing these in prior to sowing of legume seed, (3) hand planting of grass cuttings and sowing of legumes, or (4) hand planting of grass and legume cuttings.

Mac Evoy (1974), on the basis of practical experience with pastures under coconuts on Mindanao in the Philippines, strongly recommended tillage and cultivation to prepare a proper seedbed for planting pastures under coconuts. To overcome soil compaction, he recommended sub-soiling every 5 to 6 years; he further suggested that such tillage

138

Table 6.1.   The after-effect of subsoiling on coconut and pasture yields (Rajaratnam and
Santhirasegaram, 1963).

Tillage Treatment	Pasture Grasses			
	Paspalum commersonii		Brachiaria miliiformis	
	Nuts per ha	forage yield, D.M. kg/ha	nuts per ha	forage yield, D.M. kg/ha
no subsoiling	8090	1870	6410	11,610
sub-soiling	8970	2190	6400	11,730

will improve drainage as well as general thrift to
the palms.   For planting pastures the following
tillage steps were suggested;   plow twice and
harrow twice, with at least one subsoiling, if
possible.   The suggested sequence is plow, harrow,
plow, harrow, and subsoil.   First plowing should
penetrate 15 to 20 cm, second plowing (cross plowing if possible) should penetrate to 25 cm.   He
believes in plowing as deeply as possible.   Subsoiling should reach 50 cm, if possible, but should
not be done closer than 1.5 m from the base of
palms, in order to avoid root damage.

If cultivation to the base of the palms is
practiced, some root pruning will surely occur.
How much damage does this do?   The author has seen
fields in Jamaica, the Philippines, Western Samoa,
El Salvador and Sri Lanka in which this was done.
Was this injurious?   It was impossible to tell for
sure, but some roots had been severed.

Depth of cultivation would certainly be a factor.   In the Solomon Islands, cultivation to 15 cm
depth caused yield depression in coconut (Ian Freeman, Pers. Comm., 1972) however, a single cultivation to establish pastures was not considered harmful.

If the lateral roots extend out from the palm to a distance of 2m, in palms spaced 8 x 8m in a square pattern, it would appear that a 4m strip could be cultivated safely without injury to coconut roots. By cultivating in 2 directions, it should be possible to till about half of the field in order to plant pasture grasses and legumes. Such tillage could aid materially in cutting down pasture establishment costs, while at the same time providing a good seedbed for forage crops.

SCHEDULING LAND PREPARATION AND PLANTING OF PASTURES

New Coconuts

For new coconuts it would be best to plant pasture legumes or cover crops right áfter land clearing. This will assist in keeping down weed growth and will ensure a good cover. Palms could then be planted later in ring-weeded areas which should be kept clean during growth of the young plants. Legume growth could be grazed by tethering to steel pegs in the interrows of the young palms, or by hand harvesting and stall feeding of the forage. Later, when palms reach suitable size for grazing, desirable grasses can be established in the legume cover.

Established Coconuts

Cultivation during dry weather and before the onset of rains will help to ensure a good seedbed and good soil moisture for the young pasture plants. Seeds or planting materials should be on hand in order to plant as soon as land preparation is completed or when adequate moisture is available. If fertilizers or soil amendments which should be incorporated into the soil are to be used, these should be broadcast on the soil surface before the last harrowing. Phosphorus fertilizer and lime are two materials which would be of most benefit if incorporated into the soil.

PLANTING IN CIRCLES OR STRIPS

If the land must be cleared and prepared by hand, it is advisable to slash down existing vegetation and then clear and prepare planting circles one meter or so in diameter, two to four meters apart between the rows of palms (Lambert, 1970). These circles can be planted with creeping grasses or legumes. If the circles are fertilized and cared for, the vigorous grasses and legumes will spread quickly and in less than a year should reach a full stand under the palms. A modification of the circle preparation technique could be done by preparing long open planting strips; however, the spaced circles probably offer more potential for rapid, inexpensive plantings. Strip planting could be done by opening planting lines with a subsoiler or ripper or disc plow.

SEEDBEDS

When pasture grasses or legumes are established from seed, the seedbed must be carefully prepared in order to obtain a good stand. For small-seeded species, the seedbed should be fine and free of weeds. Vegetatively-propagated species usually do not require a fine seedbed, although if the plants are slow-growing in the early stages, more attention must be given to land preparation to reduce weed growth and competition during the establishment period.

ESTABLISHING THE PASTURE

Detailed instructions for individual pasture species will be presented in Chapter 8. This section will deal therefore with principles of pasture establishment.

Legumes

Most legumes can be established from seed, which can be drilled or broadcast sown (Figures 6.3, 6.4). If broadcast sown, light harrowing, raking or discing will be necessary to cover the seeds. Seeds can be covered by dragging a freshly-cut tree limb, with leaves attached, over the surface of the plowed or disced field. Most legumes will require inoculation with the right strain of Rhizobium; however, some do not require inoculation. A list of some tropical legumes and their inoculation requirements is presented in Table 6.2.

Under certain conditions a better legume stand can be obtained by seed pelleting. Reasons and directions for pelleting have been reviewed by Norris (1967) and Plucknett (1971).

Figure 6.3. Distributing guineagrass and centro seed by grain drill (above) and by hand-
operated whirling spreader-seeder (below). Joint
Research Programme, UNDP/Western Samoa Department of Agriculture, University of the South
Pacific School of Agriculture; Vailele, Western
Samoa.

Figure 6.4. A centro/guineagrass pasture sown
two months before. UNDP/Western Samoa Department of Agriculture, University of the South
Pacific School of Agriculture, Cooperative
Project, Vailele, Western Samoa.

Table 6.2. Inoculation guide for tropical pasture legumes (after Norris, 1969).

Species	Common name	Specific Rhizobium requirement?	Inoculant required
Cajanus cajan	Pigeon pea	unspecialized	None
Calopogonium mucunoides	Calopo	"	"
Centrosema pubescens	Centro	highly specialized	Centro specific
Desmodium intortum	Greenleaf desmodium	specialized	Desmodium group
Desmodium heterophyllum	--	highly specialized	D. heterophyllum species
Desmodium uncinatum	silverleaf desmodium	specialized	Desmodium group
Dolichos axillaris	Archer dolichos	unspecialized	None
Dolichos biflorus	Leichardt dolichos	"	"
Dolichos lablab	Rongai dolichos	slightly specialized	cowpea

Table 6.2. (con't)

Species	Common Name	Specific Rhizobium requirement?	Inoculant required
Glycine wightii (formerly G. javanica)	Cooper, Clarence, Tinaroo	slightly specialized	cowpea
Leucaena leucocephala	Peruvian leucaena	highly specialized	Leucaena specific
Lotononis bainesii	Miles lotononis	extremely specialized	Lotononis bainesii specific
Macroptilium atropurpureum	Siratro	unspecialized	None
Macroptilium lathyroides	Phasey bean	"	None
Pueraria phaseoloides	tropical kudzu	"	"
Stylosanthes guyanensis	Schofield stylo	"	"
Stylosanthes guyanensis	fine stem stylo	highly specialized	fine stem stylo specific

Grasses-Vegetative Propagation

Most tropical grasses do not readily produce seed, and for that reason many are vegetatively propagated by stem or stolon cuttings or by crown divisions. Freshly-cut grass materials are broadcast on the soil surface and are then disced or plowed into the soil, or are planted individually as "sprigs" in holes or furrows. A new system of pre-rooting planting materials for tropical grasses was developed by Nicholls and Plucknett (1971, 1972) in Hawaii. In this system, which has been termed "packet planting", grass cuttings are grown in small peat propagation media for 2 or 3 weeks and are then transplanted into the field in low density stands at the rate of 500 or so per ha. This method saves labor and expense in cutting and transporting planting materials, and also makes for more dependable pasture establishment.

A modification of the packet planting system can be accomplished by using coconut husks in place of the peat pots. This approach using soil in the coconut husk has been employed by WSTEC in Western Samoa (Morris Lee, Pers. Comm. 1972).

For some creeping grasses (pangola and para, for example) it is possible to mow or handcut a mature stand, gather the clipped material, broadcast it on a plowed or disced field, and then disc it in (or cover by hand immediately). Many tropical pastures are established in this way. The amounts of materials to be planted in this way vary according to the type of grass used, the amount of time allotted by the farmer for establishment of the pasture, and the age and conditions of the planting material. However, usually at least 3000 to 4000 kg per ha of fresh material will be required to get a good stand. If labor is short or if the pasture isn't needed right away, less material can be planted. Another factor to consider is that older, more mature grass cuttings are less delicate and more likely to strike and take root than young, more succulent cuttings.

Of course, the more material planted and the closer the spacing of the plants, the more rapidly the new pasture will close in, cover the ground, and suppress weed growth. Nevertheless, low density plantings of individual vegetative "sprigs" can be successful, despite the longer period of time required to cover the ground. The main principle to

use in ensuring success is to favor the desirable
pasture plant at the expense of less desirable
plants. To do this requires that fertilizers be
applied only near the desirable pasture species,
that mowing or cutting of weeds be done, etc.

Successful low density plantings of palisade
grass (Brachiaria brizantha) have been used in
Western Samoa by WSTEC. In this system one meter
diameter planting circles are cleared and prepared
and then planted with three or four sprigs of
palisade grass. A small amount of fertilizer is
applied to the planting site at planting and at
three or four month intervals thereafter during the
first year. The planting circles are spaced two to
three or more meters apart. The paddocks are
closed, and at the end of a year of undisturbed
growth, a virtually pure palisade grass pasture
is the result.

Bunch grasses (e.g., guineagrass) can also be
propagated vegetatively by using stem or crown
divisions. These can be planted in furrows or in
planting holes and will grow to produce mature
plants. However, in order to obtain a dense stand
of bunch grasses in this way, it is necessary to allow the grass to produce and shed seed. Bunch grass
establishment by vegetative means is therefore essentially a two-step process; (1) establishment of
the mature plants from crown division, and (2) establishment of a seedling population from seed.

Grasses - Establishment from Sown Seed

For grasses that produce viable seed, establishment is relatively easy and does not differ much
from legume establishment from seed. However, some
grasses have very small, light seeds that are difficult to distribute by hand or mechanically, and --
to ensure good distribution when drill or broadcast
sown -- may require mixing with sawdust or sand as
a filler.

For some grasses seedling growth may be slow
and weed competition severe.

NURSERIES TO PRODUCE PLANTING MATERIALS

Farmers should plan ahead to ensure that an
adequate supply of planting materials for vegetatively propagated grasses or legumes will be avail-

able when needed. The best way to do this is to
plant specialized areas or "nurseries" of the
desired species (Figure 6.5). The ratio of land
area devoted to nursery and land area to be planted
varies widely according to the growth rate of the
grass or legume itself, as well as the planting
system (spacing, land preparation, fertilizers used,
etc.). For example; WSTEC in Samoa plants one ha
of _Brachiaria brizantha_ for every 200 ha of pasture
to be sown, while it requires one ha of _Dicanthium
aristatum_ nursery to provide planting material for
7 ha. Of course, repeated harvests of planting material from such nurseries can be made, especially
if planting is carried on over a longer period
during the year, thus reducing the amount of land
area devoted to nurseries.

Nurseries should be located in an easily
accessible place near the area to be planted, if
possible.

SOWING PASTURE MIXTURES

There are sometimes good reasons for growing
forages in pure stands, but in most cases mixed
stands are very desirable. Some reasons for sowing
mixtures (both simple and complex) of grasses and
legumes include the following: (1) crop security
because hazards do not affect all plants equally
or at the same time, thus with a mixture a farmer
is more likely to have feed available in every
season than if he uses a single forage; (2) forage
quality is usually higher with a grass/legume mixture; (3) soils in coconut lands are often variable,
and mixed stands may furnish forage plants better
adapted to each soil type; (4) grass/legume mixtures usually resist weed encroachment better than
single stands; and (5) palatability may be higher
in a mixed stand.

Some general principles of forage mixtures include: (1) every forage mixture should include at
least one grass and one legume; (2) plants included
in a mixture should be adapted to the use(s) for
which the mixture is intended; (3) as much as possible, the mixture should be made up of plants that
mature together or at least are compatible in terms
of suitability for timely grazing, seasonal growth,
etc; (4) mixtures should not be comprised of plants
of widely-differing palatability, to reduce the
possibility that the more palatable one may be

Figure 6.5. A palisade grass (<u>Brachiaria</u>
<u>brizantha</u>) nursery, WSTEC, Western Samoa.
The palms in the background are of the
Malayan Dwarf variety.

grazed out; (5) in general, bunch grasses are usually less competitive with legumes than sod-forming or creeping grasses; and (6) sowing mixtures require that suitable amounts of each forage crop seed are sown to provide a good stand of that plant (i.e., sowing forage mixtures does not necessarily save seed).

GRAZING OF THE NEW PASTURE

All farmers must come to grips eventually with the question, when and how should a new pasture be grazed? In general, a new pasture cannot stand full grazing pressure until it is at least eight to twelve months old. Before that time, light, infrequent grazing should be done to remove tall growth of pasture plants, to allow small shaded seedlings to grow and develop, to graze off less desirable plants, and to stimulate development of lateral buds from which new growth can occur. Care should be taken to prevent overgrazing, severe trampling of young pasture plants, or soil compaction by animals on soft, recently-plowed soil.

7
Improved Pastures

Pasture management has several objectives: (1) harvesting the forage either through grazing or removal for hay, silage, or cut fodder; (2) preservation of a productive, vigorous sward; (3) maintenance of productive, desirable forage plants in the sward; and (4) provision of a continuous feed supply for animals. Management practices to obtain these objectives can include use of different stocking rates, grazing with various classes or types of animals, seasonal grazing, use of different grazing systems, use of fertilizers, selection of the proper pasture species or mixtures,weed control, fencing, location of water points or salt licks, and use of deferred grazing or reserve pastures. The management system employed will vary greatly depending on several factors, including: size of farm, skill and background of the farmer, type of soil and its fertility, rainfall, spacing of palms, cost and availability of fertilizers, availability of suitable planting materials or fencing materials, and available markets.

An added factor in managing improved pastures under coconuts is the question of the effect of grazing and pastures on coconut yields.

This chapter will discuss several types of improved pastures and their special management requirements, as well as other special problems in managing improved or established pastures under coconuts.

SEASONAL FORAGE GROWTH AND THE FEED YEAR

Seasonable growth is one problem which must be faced in most coconut/pasture/livestock systems.

Seasonal growth differences can occur for several reasons: (1) poor or irregular rainfall distribution; (2) low sunlight during rainy, overcast weather; (3) flowering or daylength responses which reduce forage yield; and (4) different growth patterns of the forage species used.

Tropical pastures also experience seasonal growth effects. For example, in Hawaii highest forage growth is expected during summer months June-September), while lowest forage is produced during winter (December-February) when about 66 per cent of summer yield can be expected. The reduction can be more pronounced with species such as pangola grass which makes little growth during winter when low temperatures are around 15ºC.

Seasonal growth differences cause problems in providing sufficient feed year-round. Often the herd size on a property is determined largely by the amount of forage available during the poorest forage production period, e.g. the winter season. This often means that for most of the year the property is understocked, that the pasture crop is under-utilized and that the farmer is not obtaining the full benefit from his crop. All farmers need to consider how to overcome seasonal forage problems and to achieve a year-round, assured feed supply. This is often referred to as the FEED YEAR. Some management decisions and adjustments could help to overcome seasonal forage growth and to achieve the concept of a FEED YEAR. Possible measures include:

(1) growing special fodder crops under very intensive management in order to provide special pasture or cut-feed for the animals during shortage periods. Such crops could include napier grass (Pennisetum purpureum), guinea grass (Panicum maximum), leucaena (Leucaena leucocephala), leucaena/guinea grass mixtures, forage sorghum, and para grass (Brachiaria mutica) (Plucknett, 1978).

(2) providing special high-producing pastures for seasonal use. Such pastures could include para grass in swampy areas, a laucaena/guinea grass mixture, and guinea grass.

152

Paltridge (1957) suggested using guinea grass as a fodder crop or as a pasture grass under rotational grazing in coconuts to provide reserve feed. Ellewela (1956b) recommended using napier grass and palisade grass (Brachiaria brizantha) in the semidry zone of Ceylon to provide fodder during the dry season when the pastures dry up.

Rajaratnam and Santhirasegaram (1963) pointed out that an adult Sinhala cow (weighing about 180 kg) can be maintained on 0.14 ha of guinea grass, while 0.44 ha of corigrass (Brachiaria miliiformis) or 0.48 ha of palisade grass were required for an animal of this size.

Regardless of what seasonal problems are faced, all cattle producers should do what they can to provide a full "feed year" for their animals. To do this may require supplementation on pastures, establishing and maintaining special pastures, growing irrigated fodder crops, or setting aside reserve pastures.

GRAZING THE PASTURE

The grazing animal is probly the most important single factor in managing pastures. Too many animals on a pasture leads to overgrazing, and consequent degradation and deterioration of the pastures. Too lenient grazing leads to under-utilization and wastage of the pasture crop. The farmer needs to find a management system for his animals that will suit his circumstances and that will match his production goals, as well as maintaining a desirable, vigorous pasture.

Stocking Rates

The number of animals grazed per hectare is the stocking rate. At low stocking rates, individual animal performance is high. As stocking rates increase, individual animal gains or production decline. At very high stocking rates, overgrazing and deterioration of the pasture may occur. Higher stocking rates usually produce higher yields than lower stocking rates, up to a point when the pasture cannot sustain very large numbers of animals.

At low stocking rates animals graze selectively, choosing the most desirable or palatable plants and

153

leaving those which are less desirable or palatable. Certain weed problems can become more serious when selective grazing occurs. As stocking rates increase, selective grazing is reduced, in part because animals are forced to eat less desirable plants. In coconut/pasture systems the need to control understory vegetation may mean that medium to heavy stocking may be required to obtain the dual benefits of close grazing to expose fallen nuts and to reduce pasture competition with coconuts. Under these conditions individual animal performance is likely to be reduced. When "brushing" or weed control is the primary purpose for grazing, individual animals may gain or produce very little.

Choice of stocking rate depends on the type of pasture, its potential production, and possible alternative feed sources, as well as the total number of animals which the farmer must provide for. When stocking rates are too high, only a few alternatives are possible; some animals must be removed and placed on pasture elsewhere (not a very viable solution for most small farmers), supplemental feeding on pasture can be done, feeds can be purchased, fodder from roadsides or waste areas outside the farm may be cut and fed, or some animals must be sold.

The most suitable stocking rate to use would be that one which: (1) does not damage or degrade the pasture sward; (2) allows stability of the system and will not place undue pressure on the farmers to purchase high-priced feeds, to sell animals during periods of drought or feed shortage, or to expend extra labor or expense on cutting and hauling feeds; and (3) which will provide marketable or saleable livestock products (Humphreys, 1974). To satisfy these requirements usually means that stocking rates will be somewhat below the maximum potential production of the pasture. Often, that period of the year during which least forage is produced sets the stocking level for the farm. This means, however, that for most of the rest of the year pastures will be underutilized. Potential stocking rates for various pastures will be presented in the discussions of individual pastures.

Grazing Systems

Grazing systems are discussed in detail in

Chapter 9; however, there are some points which should be discussed here.

Continuous grazing is widely practiced and is a quite satisfactory system in many ways. The amount of fencing required is less than for other systems. With lenient grazing certain desirable legumes persist and remain productive, and individual animal performance is usually higher than for other grazing systems. On the other hand, selective grazing often occurs, and buildup of less desirable species can result. Recent research indicates that under the shaded conditions of the coconut/pasture system, continuous grazing may be less suitable than rotational grazing, particularly for less shade tolerant grasses (Whiteman, 1977).

Rotational grazing is well suited for coconut/pasture systems because it fits directly into harvesting and gathering schedules for the nuts. Monthly nut collections scheduled just at the end of a grazing period allow near-perfect nut harvests. Rotational grazing usually results in less selective grazing, and it also provides a period of rest and recovery for grasses and legumes that are injured by heavy, continuous grazing, especially for those plants which are constantly selected by the grazing animal.

Deferred grazing is probably not too well-suited to coconut/pasture systems because of the need to reduce understory competition with the coconuts. Hence, deferred or reserve pastures should be grown outside the coconut groves, if possible. If it is not possible, then perhaps areas with older, more widely-spaced palms should be selected for deferred rotational grazing or as reserve pastures.

Zero grazing (cut-and-carry feeding, soilage cropping) is a way to increase forage production on a farm, while at the same time meeting requirements such as labor availability and utilization, as well as reducing forage waste or spoilage.

Zero grazing systems can be combined with grazing in order to provide feed throughout the year and to allow the farmer to carry more animals on the farm.

Zero grazing allows a farmer to intensify forage production on his better lands, while using his poorer lands for pasture. It also provides a means to produce manure for palms and food crops.

Some farmers may decide to use zero grazing as their sole management system. It allows reduced fencing and increased forage utilization, and if practiced properly animal performance can be good to excellent. It is especially suited to dairy production or finishing of beef animals.

Supplementation on pasture can be used to help overcome periods of drought or feed shortage. Supplemental feeds could be crop residues such as rice straw, corn or sorghum stover, sugarcane leaves, etc., or industrial by-products such as fish meal, copra meal (known as "poonac" in Sri Lanka), and molasses.

One caution should be given; supplemental feeding may reduce pasture utilization, because animals may just substitute the one feed for the other.

Grazing with different classes of animals

Animals vary widely in their grazing habits and preferences. This is true both within groups such as cattle where age, sex, and other factors are important, as well as between species such as sheep, goats, domestic buffalo, or cattle.

In some countries different groups of animals may be grazed together to obtain better utilization of the available forage (see Chapter 9). For example, cattle and sheep are sometimes grazed together because each may graze different plants in the pasture. Also, steers may be used to utilize lower quality forages, while better quality pastures are reserved for use by young weaner stock or bred heifers. Also, cows with calves at side or dry cows can utilize poorer quality forage than young stock being fattened for marked, or dairy cows in full production.

MANAGING THE PASTURE

A major goal in pasture management should be to obtain high forage and animal production, while at the same time maintaining a desirable, vigorous sward.

What is the most desirable pasture for use in
coconuts? Probably the most important characteristics should be: (1) little competition for nutrients
and moisture with coconuts; (2) shade tolerance; (3)
low stature, or ability to withstand low grazing so
nuts can be seen on the ground; and (4) ability to
grow in productive, compatible grass-legume mixtures.

NUTRITIONAL REQUIREMENTS OF PASTURE SPECIES

Grasses and legumes vary widely in their nutritional requirements. Also, many species have rather
specific requirements. A full discussion for each
of these species is outside the scope of this book;
however, an attempt will be made to point out those
nutritional requirements of pasture plants which will
either complement or cause difficulty in providing
adequate nutrition for coconuts when pastures are
grown underneath, as well as in affecting pasture
performance and yield.

For grasses, the most widespread nutrient
deficiency is nitrogen (N). For pure grass pastures
N must be supplied by fertilization. When grass
pastures and coconuts are grown together, the main
consideration will be applying adequate fertilizer N
to avoid yield losses in coconut because of competition from the grasses. When high yields of forages
are required in intensive forage production systems,
it may be necessary to use commercial N fertilizers
and not to rely on legumes to supply N for the
grasses.

Most grasses do not have a high phosphorus (P)
requirement. Provided enough is applied to take
care of both grasses and coconuts, little competition for this element should occur in pure grass
pastures. As with P, most grasses do respond to
potassium (K) fertilization on deficient soils. The
K requirements of most tropical grasses are modest,
but because of the high K requirement of coconut, K
fertilization must be practiced carefully in coconut/
pasture associations. Care should be taken to ensure
that the grasses do not compete severely with coconut
for K.

Pure or mixed legume stands will require P and
K fertilizers on some soils. In many tropical soils,
limited P availability may be the primary limiting

157

nutritional factor in legume establishment, yield, and N fixation, whether legumes are grown alone or in mixtures with pasture grasses. For best results, P fertilizers should be broadcast and incorporated into the soil before planting. In leached acid soils, calcium (Ca) and magnesium (Mg) may be needed to obtain good legume growth. Well-nodulated legumes should produce good N yields through symbiotic N-fixation, and some of this N should be available for the coconut palms. Whether the legume can supply enough N for optimum coconut yields is not known; however, higher nut yields are known to result with legume cover crops than when non-legume cover crops are used. Grazing of legume cover crops could result in shedding of nodules from the legume roots, thereby releasing N for the coconuts. When legumes which fix small quantities of N are used as cover or pasture crops, supplemental N fertilization of coconuts may be required.

Fertilizing the pasture

Fertilizer use will greatly affect pasture production. For grass pastures, N will be the most important limiting nutrient. For legume/grass pastures, provided that the legume can produce enough N for the grass, then P and K are likely to be most limiting. Pastures are most likely to compete with coconuts for N and K, and for that reason alone fertilization becomes extremely important in this farming system.

Timing of fertilizer use is very important in pastures. Fertilizers should be applied during periods of good rainfall in order to ensure that full benefits can be obtained. Strategically-timed fertilization of pastures just before the dry season can help to provide extra forage during difficult times when forage production is drastically curtailed.

Fertilizers are important tools in balancing legume and grass growth in legume-based pastures. Use of N fertilizers, for example, on legume-based pastures can greatly stimulate the more vigorous grasses and lead to a weakening or even disappearance of legumes from the sward. Conversely, fertilizers which stimulate legumes can help greatly in improving their competitiveness and persistence.

Fertilizer placement is important in coconut/
pasture systems, for it will affect the ability of
palms to obtain required nutrients without serious
competition with the pasture. Also, many countries
recommend that fertilizers for coconuts be buried or
incorporated into the soil. In coconut/pasture
systems, this would be very difficult. In order to
evaluate this, Santhirasegaram (1959) used four
methods of fertilizer application in two types of
pasture (Table 7.1). Highest nut yields over a
four-year period were obtained by broadcasting fertilizer in a circle around each palm. This method
lends itself well to coconut/pasture systems.

Table 7.1. Comparative nut yields (in nuts/plot)
of two grass/coconut pastures, with four methods
of fertilizing the palms (after Santhirasegaram,
1959). 1/

Pasture grass	Buried in trenches	Buried in circles	Broadcast in strips	Broadcast in circles
Paspalum commersonii	260	285	336	355
Brachiaria miliiformis (cori grass)	273	233	234	293
Total	535	519	571	643

1/ Yields are an average of 4 years (1957-60).
   Fertilizer treatment on the palms was 4.5 kg
   of an NPK fertilizer (1:1:1 mixture) per
   palm every 2 years. Fertilizers for the
   pastures were 54 kg N/ha as ammonium sulfate broadcast in split applications in
   January, May, July, and October and 37 kg
   K/ha as muriate of potash in July.

EFFECT OF GRAZING AND PASTURES ON COCONUT YIELDS

The only critical work on this subject has been done in Sri Lanka and Western Samoa, where the effects of pasture species, fertilization, and grazing intensity have been studied. The following section will present a discussion of some of the research in Sri Lanka and its implications for managing pastures under coconuts.

In the early work, the effect of four grass pastures on coconut yield was compared. Paspalum commersonii provided fewest grazing days per ha per year and also produced lowest nut yields (Table 7.2). Palisade grass and cori grass were about equal in grazing days and nut yield, while guinea grass provided the highest number of grazing days and a slightly lower nut yield than the two Brachiaria species.

Later studies indicated that palisade grass pastures tend to depress nut and copra yields as compared to other pasture species (Tables 7.3, 7.4, 7.5, and 7.6). In some cases nut yields from palisade grass pastures were lower than yields from ungrazed natural covers (weeds). It should be remembered, however, that palisade grass was also providing feed for cattle during this period while weeds would be of little value as forage.

Cori grass does not depress nut and copra yields to the same degree as palisade grass (Tables 7.3, 7.4, 7.5, and 7.6). For this and other reasons cori grass is highly favored in Sri Lanka for pasture under coconuts. Before leaving this subject it should be made clear that in most of these studies coconut yield from palisade grass and cori grass pastures were not much lower than the natural (weedy) cover which would be present regardless of whether pasture improvement was practiced or not.

Ungrazed grasses apparently cause a more depressive effect on nut and copra yields than do grazed grasses, and this seems to be especially true of palisade grass (Tables 7.7, 7.8, and 7.9). Over a 3-year period, ungrazed palisade grass yielded 10,344 nuts/ha/year; however, stocking at 0.4 ha/animal, 0.6 ha/animal, or 0.8 ha/animal produced average nut yields of 11,476, 10,547, and 12,562 nuts per ha per year, respectively. In later stud-

Table 7.2. Nut yields of coconuts with grazed underplanted pastures (after Santhirasegaram, 1959).

Pasture grass	Number of grazing days 1959	Number of nuts/ hectare [1] Average, 1956-60
Paspalum commersonii [2] (control)	284	7,320
Brachiaria brizantha [2]	450	8,420
Brachiaria miliiformis [2]	408	8,600
Panicum maximum [3]	675	8,220

[1] 158 palms per hectare, 8.5 x 8.5 m spacing.

[2] 0.4 ha paddocks, 28 days of grazing maximum per plot.

[3] A 0.4 ha pasture subdivided into eight small paddocks, grazed rotationally by 9 cattle.

Table 7.3. Effect of two grass pastures on nut yields per palm (after Ramalingam, 1961). [1]

N-P-K fertilizer kg/ha	Brachiaria miliiformis nuts/palm	Brachiaria brizantha nuts/palm	Control (weeds) nuts/palm
250-125-250	45.5	37.5	35.5

[1] Values are a mean of 1959 and 1960 yields. All pastures (Brachiaria spp.) were lightly harrowed before fertilization; control was not harrowed.

Table 7.4. Effect of pasture species on nut yield of coconut (After Santhirasegaram, 1964).

Pasture or cover	Number of nuts per hectare	copra yield kg/ha
Weeds	8,319	1,956
Brachiaria brizantha	6,491	1,480
Brachiaria miliiformis	7,659	1,807
Least significant difference (0.05)	864	129

Table 7.5. Effect of pastures and nitrogen on nut and copra yield. (After Santhirasegaram, 1965).

Pasture species	Nut yield/ha			Copra yield, kg/ha		
	Nitrogen/tree/kg			Nitrogen/tree/kg		
	0.33	0.65	1.31	0.72	1.44	2.88
Weeds	3218	3957	3853	1933	2043	1899
Brachiaria miliiformis	3066	3135	3582	1614	1641	1805
Brachiaria brizantha	2730	2634	2349	1389	1378	1655

Table 7.6. Effect of pastures and potassium fertilizer on nut and copra yield (After Santhirasegaram, 1965).

Pasture or cover	Number of nuts/ hectare		Copra, kg/ha	
	Potassium fertilizer, kg/tree		Potassium fertilizer, kg/tree	
	K_1	K_2	K_1	K_2
Weeds	8,739	10,265	1,987	2,400
Brachiaria miliiformis	7,859	8,252	1,828	1,951
Brachiaria brizantha	6,866	7,319	1,572	1,729

Table 7.7. The effect of intensity of grazing and level of fertilizer on yield of coconut (after Santhirasegaram, 1976b; Ferdinandez, 1968, 1969, 1970, 1971).

Cover/pasture	fertilizer [1] treatment	grazing [2] intensity	Brachiaria brizantha 1966 - 1970 Average		Brachiaria miliiformis 1973 [3]	
			No. of nuts/ha	forage dry matter, g/m^2	No. of nuts/ha	forage dry matter, g/m^2
Weeds	N	nil	10,756	181	14,912	208
Brachiaria brizantha	N	nil	5,643	376	8,036	417
"	N	N	9,596	372	11,162	480
"	N	H	7,490	281	8,391	350
"	H	N	10,108	366	11,120	343
"	H	H	11,601	190	14,757	496

N = Normal fertilizer; = 28 kg N, 9 kg P, 63 kg K per ha per year.
H = High fertilizer; = 56 kg N, 18 kg P, 125 kg K per ha per year.
2) N = 5 adult Sinhala cows/ha (an adult cow weighs about 180 kg)(Goonesekera, 1967).
H = 39 adult Sinhala cows per ha.
3) In late 1970, Brachiaria miliiformis was substituted for Brachiaria brizantha

Table 7.8. The effect of two pasture grasses and four grazing intensities on coconut yields (nuts per hectare). (After Santhirasegaram, 1959). [1]

Pasture or cover	Nut yield per hectare Average, 1957 - 1959
No grass (weeds)	12,804
Paspalum commersonii (not grazed)	12,219
Paspalum, 0.4 ha/ animal	12,330
Paspalum, 0.6 ha/ animal	10,962
Paspalum, 0.8 ha/ animal	12,557
Brachiaria brizantha (not grazed)	10,344
Brachiaria, 0.4 ha/ animal	11,476
Brachiaria, 0.6 ha/ animal	10,549
Brachiaria, 0.8 ha/ animal	12,562

[1] At planting, pastures received 28 kg P/ha and 280 kg K/ha. After planting, N was applied at 25 kg N/ha every 3 months. Four months after planting 560 kg of lime/ha was applied. Palm fertilizers were not specified.

Table 7.9. Effect of grazing intensity and
Brachiaria brizantha pastures on nut yield in
coconut (After Rajaratnam and Santhirasegaram,
1963).[1]

Cover or pasture	Grazing intensity	Nut yield/ hectare/year
Weed cover	ungrazed	13,684
Brachiaria brizantha	ungrazed	10,028
Brachiaria brizantha	2.5 animals/ ha	11,411
Brachiaria brizantha	1.25 animals/ ha	11,231

[1] Grazing was continued for 3 months, with 3
   months rest.

Table 7.10. Nut yield from coconuts under
which cattle were grazed on palisade grass
pastures using rotational and continuous grazing systems (after Ramalingam, 1961).[1]

Grazing system	Number of nuts per ha per year (1960)
Continuous grazing	10,000
rotational grazing	9,865

[1] Fertilizers applied were 25 kg N per ha in
   May; 25 kg N, 63 kg K and 11 kg/ha in
   October.

ies (Table 7.7) fertilized but ungrazed palisade grass produced an average of 5,782 nuts/ha/year, while fertilized palisade grass pastures stocked at 5 animals per ha and 40 animals (Sinhala cattle) per ha produced 10,179 and 7,909 nuts/ha, respectively. If one animal unit (AU) is considered to be a 1,000 pound animal per acre (= 1,121 kg/ha), 5 Sinhala cows per ha are equal to 0.8 AU, while 40 animals/ha are equal to 6.42 AU per ha.

Santhirasegaram (1967b) and Ferdinandez (1968, 1969, 1970, 1971) studied the effect of grazing intensity and levels of fertilizers on coconut yield in Sri Lanka (Table 7.7). They compared an ungrazed weed cover, ungrazed palisade grass,"high" and "normal" levels of fertilizers, and "high" and "normal" grazing intensities. "Normal" grazing intensity was five adult Sinhala cows (about 180 kg liveweight each) per ha. Highest nut yields were obtained with fertilization and the high stocking rate. In decreasing order, nut yields were; high fertilizer/ high stocking rate > ungrazed weed cover > high fertilizer/normal stocking rate > normal fertilizer/ normal stocking rate > normal fertilizer/high stocking rate > ungrazed, fertilized (normal) palisade grass. Forage yields decreased in the following order; ungrazed, fertilized palisade grass > high fertilizer/normal stocking rate > normal fertilizer/ normal stocking rate > normal fertilizer/high stocking rate > high fertilizer/high stocking rate > ungrazed weed cover.

Santhirasegaram (1959) studied nut yields from two grasses and four levels of grazing intensity (Table 7.8). Highest nut yields were obtained from a natural cover and from Paspalum commersonii pasture stocked at 0.8 ha per animal. Ungrazed palisade grass produced lowest nut yields.

Increasing fertilizers on pastures can reduce nut losses due to competition from the grasses (Table 7.5). Increasing N on cori grass pastures produced both higher nut and copra yields. In palisade grass pastures, while nut yields declined slightly with increasing N application, copra yield per ha was highest at the highest rate of N applied per palm, indicating that although fewer nuts were harvested, the nuts were larger. Increasing K fertilization of pastures also raised nut yields (Table 7.6). Table 7.7 shows the effect of increasing fertilizer and grazing intensity on nut yields. In this study highest nut yields were obtained by high fertilization and the highest stocking rate. Results of a comprehensive study of NPK fertilization on nut yields of palisade grass pastures are presented in Table 7.11. The data clearly show the increased nut yields which result from heavy fertilization.

This research work indicates that pasture use will probably not have depressive effect on nut and copra yields, provided that fertilizers are applied to the pastures and the palms. Also, grazing of pasture species, especially palisade grass, will reduce ground cover and resultant competition for moisture for the palms.

The effect of rotational versus continuous grazing on nut yield was investigated by Ramalingam (1961). There appeared to be little difference in nut yields between the two systems (Table 7.10). It is probable that provided the grazing system employed keeps grass competition with coconuts to a minimum, the type of grazing system will have little effect on nut yields. Of these two systems, however, very lenient continuous grazing would be most likely to leave too much grass to compete with coconuts, as well as shielding fallen nuts from view at harvest.

IMPROVED PASTURES - MAJOR SPECIES OF GRASSES AND MIXTURES

Shade tolerance of pasture plants

Recent research has yielded very interesting results as to relative shade tolerance of forage grasses and legumes. Ericksen (1977) studied forage yields of 6 grasses and 6 legumes at 100, 70, 45, and 27 per cent(%) daylight using polypropylene netting. Nitrogen content of grasses increased with shading. The most shade tolerant grasses were guinea, cori, and signal. Under 27% daylight yields were between 8 and 15 mt/ha/yr. When no nitrogen was applied, the highest yields were obtained at 45% and 27% daylight. Legume tolerance to shading (in decreasing order) was: greenleaf desmodium, centro, kaimi clover (Desmodium canum), leucaena, siratro, and stylo.

Table 7.11. The effect of level of fertilizer (N P K) on a coconut/pasture [1] association (After Santhirasegaram, 1966a, 1967b; Ferdinandez, 1968, 1969, 1970).

Fertilizer treatments, kg/ha	Average, 1965 - 1969	
	No. nuts/ ha	forage dry matter, g/m^2
1. N54 P9 K63	10,851	364
2. N108 P9 K63	10,090	421
3. N54 P9 K125	11,782	370
4. N108 P9 K125	10,490	424
5. N54 P18 K63	11,708	314
6. N108 P18 K63	10,840	339
7. N54 P18 K125	12,856	376
8. N108 P18 K125	11,095	370

1) Brachiaria brizantha pasture under coconuts. All fertilizers applied broadcast. Pastures grazed by 5 adult Sinhala animals per ha (about 180 kg liveweight x 5 = 2 animal units/ha).

Studies have also been conducted in the Solomon Islands using 30, 50, and 70 per cent sunlight (Watson, 1977). Low shade tolerance was found in para grass and stylo cv Endeavour. Shade tolerant grasses included cori, ruzi, guinea var. Makueni, and guinea cv Petrie Green Panic.

Brachiaria miliiformis (cori grass)

Following the discovery by the Coconut Research Institute of Sri Lanka that cori grass was an outstanding pasture grass under coconuts (Figure 7.1), this plant has been introduced and is now under study in a number of coconut-growing countries. It establishes easily and quickly, is very shade tolerant, and does not compete much with coconuts.

Management. Because of its shade tolerance, cori grass is more likely to be successful under young palms or in more closely-spaced groves. It can be established easily from seed or cuttings. The plant grows rapidly, and -- when well established -- is ready for grazing in about 45 days (6 weeks) when the crop is about 35 to 45 cm high (de Guzman, 1974). It should be grazed down to a height of about 8 cm.

Cori responds readily to fertilization, especially N, in the shaded coconut groves, and N fertilizers cause it to be very competitive with weeds. If fertilized, cori does not compete much with coconuts (see Tables 7.2, 7.3, 7.4, 7.5, and 7.6). For example, over a 14 year period in Sri Lanka, nut yields were higher when coconut was intercropped with cori (10,183 nuts per ha per year, average) than when intercropped with palisade grass (9,504 nuts/ha/yr), a natural cover (weed) control (8,448 nuts/ha/yr), or guinea grass (5,376 nuts/ha/yr).

The fertilizer recommendation for cori in Sri Lanka is 50 kg N, 25 to 30 kg P and 60 to 65 kg K per ha, split in 2 applications each at the start of each monsoon season (Appadurai, 1968).

Productivity of the pasture. Annual forage yields of cori grass average around 9,000 kg per ha (Appadurai, 1968; de Guzman, 1974). With adequate fertilization, carrying capacity is about 1.6 adult animals per ha. Under conditions of well-distributed rainfall, year-round grazing is possible. Crude

170

Figure 7.1. A cori grass (<u>Brachiaria</u> <u>miliiformis</u> pasture in Sri Lanka.

Figure 7.2. <u>Brachiaria</u> <u>brizantha</u> planted in 4-year-old underplanted palms. Underplanted areas could be used as a nursery to provide planting material of improved pasture species for pasture improvement programs.

protein yields of cori are about 8 to 10% (dry matter basis) for fertilized grass at 4 weeks, and 7 to 9% at 8 weeks (Ferdinandez, 1972b). Animals prefer cori to palisade grass (Ferdinandez, 1973).

At Lunuwila in Sri Lanka it has been possible to obtain more than 1,970 cow days (Sinhala X Sindhi crossbred cows, weighing about 270 kg each) per ha on a cori grass pasture fertilized with 67 kg N/ha/yr.

Compatibility with Legumes. With its creeping growth habit cori should be quite suitable for mixing with legumes. Only in Sri Lanka has much work been done on studying cori/legume mixtures. Puero, calopo, and centro were evaluated, and only centro could persist to any extent under grazing (Ferdinandez, 1973). Evaluation of legumes with cori should be an essential research priority.

Special problems or strengths. The major strength of cori is shade tolerance. However, it also can withstand drought and grazing as well as palisade grass, particularly under shade. Ericksen (1977) and Ericksen and Whitney (1977) have confirmed the shade tolerance of cori grass, and have found it yields better in shade than in the open.

Brachiaria brizantha (palisade grass) and Brachiaria decumbens (signal grass)

Palisade is one of the major pasture grasses used under coconuts; is very important in Sri Lanka and Western Samoa (Figure 7.2); and is being evaluated for coconut pastures in Thailand (Shelton, 1977), the Solomon Islands and the New Hebrides. A hardy competitor with weeds in Western Samoa, it forms virtually pure stands within a year from planting. More competitive with coconuts than its relative, cori grass, it is nonetheless a very valuable coconut pasture plant. Its relative, signal grass, which it closely resembles and with which it is frequently confused, has been tested successfully in the Solomon Islands under coconuts (Gutteridge and Whiteman, 1978), and is considered to be a potentially valuable pasture crop under coconuts.

Management. To prevent competition with coconut, palisade should be grazed down to a height of about 8 to 12 cm on a rotational basis, and palms

172

must be fertilized. Palisade paddocks can be grazed every 4 to 6 weeks, according to work in Sri
Lanka. The fertilizer schedule in Sri Lanka is the
same as for cori grass; 50 kg N, 25 to 30 kg P and
60 to 65 kg K per ha, split in 2 applications at the
start of each monsoon season.

Palisade can withstand heavy grazing under a
rotational system. It is adapted to a wide range
of soil conditions, but does require better-drained
soils than para grass. Its creeping habit, shade
tolerance, resistance to trampling, and rapid recovery after grazing make it especially well-suited
to coconut pastures. Signal produces some viable
seed, and commercial seed is available from Australia.

Productivity of the pasture. Both palisade and
signal are high-yielding. In the wet tropical
coastal areas of Australia, signal has produced
highest yields of dry matter -- 36 metric tons/ha/yr
-- and annual liveweight gains of over 1,000 kg/ha
when fertilized with 195 kg N/ha/yr (Humphreys,
1974).

Annual dry matter yields of palisade under
coconuts are in the order of about 8,000 to 9,000
kg/ha. In Western Samoa, palisade pastures have the
highest calving percentage of all WSTEC pastures
(Reynolds, 1976b), and a fertilized palisade pasture
produced 330 to 370 kg liveweight gain/ha/yr, 11,000
to 17,000 kg dry matter/ha/yr with an average 10 to
12% crude protein content, 5,400 to 6,170 nuts/ha,
and 1,060 to 1,170 kg copra/ha/yr (Reynolds, 1977d).

Compatibility with Legumes. Because of their
aggressive creeping habit, neither palisade nor
signal mix very well with legumes in unshaded areas.
However, in the shade of palms the grasses lose some
of their dominance, and shade-tolerant legumes that
can withstand medium to heavy grazing combine well
with them. Centro and puero are examples of legumes
that do fairly well in such conditions (Figure 7.3).
Greenleaf desmodium and siratro (in less-shaded
situations) are other examples of legumes that might
perform well with these aggressive grasses.

On a coral sand soil in the Solomons, Gutteridge
and Whiteman (1978) recommended that a signal/siratro
mixture be studied for possible use.

Figure 7.3. Palisade grass (<u>Brachiaria</u> <u>brizantha</u>) and tropical kudzu (<u>Pueraria</u> <u>phaseoloides</u> in 4-year-old underplanted palms.

Figure 7.4. A para grass (<u>Brachiaria</u> <u>mutica</u>) pasture in Jamaica.

Special Strengths. Both palisade and signal can thrive and become dominant under low fertility soils and a wide variety of sites. Their shade tolerance is well above average, and both are high-yielding.

Brachiaria mutica (para grass)

Para is moderately shade tolerant and therefore may become a major component of pastures under coconuts (Figure 7.4). In the Philippines, para/centro pastures can be very productive without competing with the palms (Philippine Council for Agriculture and Resources Research, 1976c). Para is one of three grasses recommended for use under coconut in the Philippines and is under study for use in Thailand (Boonlinkajorn and Duriyaprapan, 1977).

Management. Para can be grazed or harvested as a soilage crop under a cut-and-carry system. The grass can be harvested or grazed every 30 to 60 days. To ensure good recovery, para should not be grazed down as short as palisade or cori; it would be best to graze it down to 15 to 20 cm or so, and then move the animals to allow the grass to regrow. Provided overgrazing or excessive defoliation is not practiced, para pastures can be stocked fairly heavily, about 2 animals per ha in coconut lands. Para is not as shade tolerant as palisade or signal.

Para responds readily to fertilization. It can be quite productive, although it is not usually as productive as signal or guinea. In Thailand, a stand of para/centro cut every 45 days produced 60 t fresh forage/ha/yr; of this yield 77% was para grass (Allo, 1976). Dry matter yields under cut-and-carry systems should range from 12 to 18 t/ha/yr.

Compatibility with Legumes. Para can grow fairly tall, up to 1.5 m or more, but when it is grazed well it does not shade out legumes grown in mixtures with it. Legumes that combine well with it include siratro, hetero, centro, and -- in poorly drained areas -- phasey bean (Macroptilium lathyroides).

Special Problems or Strengths. Para is easily established, fits into poorly drained lands, and can be highly productive. It can become a weed in many plantation crops.

175

Brachiaria humidicola (syn. B. dictyoneura)
(Koronivia grass)

Koronivia is used as a pasture plant in the
South Pacific, particularly in Fiji, but is being
evaluated for use under coconuts in the New Hebrides,
Solomon Islands, and Western Samoa. It appears to
have some shade tolerance, produces well during
winter, and is well adapted to coral sandy soils
(Ranacou, 1972; Gutteridge and Whiteman, 1978).
Lever's Plantations in the Solomon Islands consider
koronivia to be the best grass for low coral sand
soils.

Koronivia mixes well with siratro and gylcine
(Glycine wightii). Reynolds (1977d) mentions that
-- contrary to local opinion -- it may be quite
palatable for livestock when grown under coconut.
In Fiji it can carry 2.5 animals per ha during winter, a time when many other grasses cannot sustain
such a carrying capacity.

Dr. Brian Robinson (pers. comm., 1972) considers that koronivia is the best grass under coconuts
in Fiji because it; (1) is shade tolerant, (2) is an
easy pasture in which to collect nuts, (3) does well
in better-drained wet soils, (4) makes good winter
growth, and (5) does well during the dry (winter)
season. Koronivia does produce a small amount of
seed in Fiji, but it is usually vegetatively propagated.

Panicum maximum (guinea grass)

Many forms and varieties of guinea can be used
under coconut; however, in general it should not be
the grass of choice for coconut pastures because it
does compete significantly with the palms. Also,
its upright, bunch-type growth habit causes nut
collection to be difficult (Figure 7.5). Nonetheless, its shade tolerance and high yields -- coupled
with drought resistance and tolerance of less fertile soils -- have caused it to be used under coconuts in a number of countries.

Of the several varieties and types of guinea
available, two may be most suitable for coconut
pastures. These are green panic (P. maximum var.
trichoglume) and "creeping guinea" (e.g. cv "Embu"
from Australia) which has a more prostrate growth

Figure 7.5. Above: Newly established guinea grass/centro pasture, just before grazing; Vailele, Western Samoa. Below: A guinea grass pasture in Sri Lanka.

habit (Weightman, 1977).

Management. Two critical practices, fertilization and grazing management, are important in using guinea under coconuts. Fertilization is required to reduce competition for nutrients between the grass and the palms. Careful grazing management will ensure that the grass is grazed down sufficiently to prevent competition with the palms and to allow for nut collection. However, too heavy grazing can lead to injury to the growing points of the grass, and eventual death of the grass may result. A rule of thumb is to graze it to a height of about 15 to 20 cm; some giant cultivars may need to be grazed no lower than 30 cm. However, it may not be easy to keep guinea grazed as low as 15 to 20 cm, because of the coarse nature of the lower stems. Rotational grazing can be of value in keeping guinea at a suitable height, as can use of smaller paddocks and higher stocking rates under rotational grazing.

Green panic is smaller than common guinea and may be less shade tolerant, but it is also less competitive with the palms. In older, wider-spaced coconuts, green panic may be a satisfactory pasture grass.

Guinea can be used as a cut-and-carry forage or as a green chop feed.

Productivity of the Pasture. Guinea pastures can be high-producing, and often carry more than 2 to 2.5 animals per ha. The grass does well under lower rainfall than many other species, and for that reason often provides year-round grazing where other grasses would be more seasonal.

Annual dry matter yields under coconuts are in the range of 10 to 18 tons per ha. In a 15-month grazing period in Jamaica, guinea was grazed 9 times and produced 605 cow grazing days per ha, a carrying capacity of 1.65 cows per ha, 126 t/ha of fresh forage, and 18 t/ha of dry matter (Coconut Research Board, 1971). Over a 2½ year period in Western Samoa, a guinea/centro pasture yielded 380 kg liveweight gain/ha/yr as compared with a good native pasture that yielded 200 kg liveweight gain/ha/yr; this is a difference of 87.5 per cent (Reynolds, 1977d).

178

Compatibility with Legumes. Guinea mixes especially well with centro, and this is particularly
fortunate, for centro is a major legume for use
under coconuts. As a bunch grass, other climbing or
creeping legumes (e.g. siratro, glycine, and greenleaf desmodium) can become established in the open
ground between the bunch grass clumps and climb up
on the grass.

Special Problems. As has been mentioned, competition with palms is the major difficulty with
guinea. "Creeping guinea" and green panic should be
evaluated more fully for use in coconut pastures.

Pennisetum purpureum (napier grass, elephant grass)

Napier is not really suited for use under coconuts (Figure 7.6). True, it is somewhat shade
tolerant and is very high-producing, but it competes
severely with coconuts for nutrients (especially for
potassium) and water, and it grows much too tall to
allow good nut collection. It is much better
suited to intensive production in pure stands in
open lands, even small parcels, where it can be
managed as a soilage (cut-and-carry) crop. Under
these conditions it is unmatched.

Management. Grazing of napier is difficult
under almost any situation. The grass is so tall
and often becomes so coarse and woody that it is
very difficult to keep down to a manageable level.
In any case both the grass and the palms should be
fertilized (particularly with N and K) to prevent
severe nutrient competition with the palms.

In most cases napier should be managed as a
soilage crop and as a cut-and-carry feed. Under this
system production is very high, the height of the
crop can be managed, and forage losses will be reduced. It can be cut at a height of about 15 to 20
cm (for some very large cultivars a 30 cm cutting
height may be required) and still leave sufficient
stem to provide buds for regrowth. Harvested every
30 to 60 days as a cut forage, napier can be high-
yielding while at the same time producing good
quality forage. If grazed or cut every 30 days just
before nut collection, napier can be managed successfully under coconuts, provided fertilization is
also practiced.

179

Figure 7.6. <u>Above</u>: A napier grass (<u>Pennisetum</u> <u>purpureum</u>) pasture with underplanted 3-year-old palms; Mulifanua Plantation (WSTEC), Western Samoa. <u>Below</u>: Napier under palms about 13 to 20 years old. Nut collection would be very difficult in such pastures. WSTEC, Western Samoa.

Productivity of the Pasture. In the open,
napier can produce 200 t or more of fresh forage/ha
(60 t/ha of dry matter) per year. Under the shade
of coconuts this yield will be reduced, but just how
much the reduction will be varies greatly with the
conditions at the time. Under cut-and-carry feeding
systems, there should be little doubt that napier
could produce enough forage on one hectare to feed
8 to 10 mature animals. In such systems if leucaena
(ipil ipil) is also fed along with the napier to
increase the protein supply, animals should gain and
produce very well.

Compatibility with Legumes. Napier is not
really compatible with legumes because of its height
and competitiveness.

Special Strengths or Problems. Its special
strengths are high yield, adaptation to less fertile
soil, and hardiness. When over-mature, its forage
quality is low. It is not well-suited for most
grazing systems, and can be very competitive with
coconuts.

Dicanthium aristatum (Alabang X, Angleton grass)

Alabang X is a Philippine selection that has
become popular under coconuts in that country. Somewhat slow to become established and susceptible to
weed competition when young, it is an excellent grass
because it: (1) is very palatable, (2) mixes well
with legumes, (3) is fairly drought resistant, (4)
withstands close grazing, (5) can be vegetatively
propagated, (6) produces high animal production, and
(7) can be used for green chop or as cut-and-carry
feed.

Management. First-year management of Alabang X
is very important, and efforts must be taken to ensure that the pasture is not overgrazed or abused
during this period when the grass is rather slow-
growing and susceptible to weed competition. Only
light grazing should be practiced during this time,
perhaps 4 or 5 in the first year are possible
(Mac Evoy, 1974). The pasture should be fertilized
after the first grazing.

de Guzman and Allo (1975) suggest that Alabang
X is best suited for lower fertility soils, and that
on more fertile sites other more vigorous plants tend

to dominate it. In less fertile soils in the Philippines it has been able to compete with imperata.

Productivity of the Pasture. Mac Evoy (1974) reported that on commercial properties on the island of Mindanao in the Philippines Alabang X fed as green chopped feed produced higher cattle live-weight gains than para or guinea grasses. Also, Alabang X yielded about 30 per cent less green chop than para grass under the same conditions.

Compatibility with Legumes. Alabang X is a low, creeping grass that is not exceptionally competi-tive; therefore, it mixes well with a number of legumes.

Special Strengths or Problems. Alabang X is resistant to drought, does not stand waterlogging, is very palatable, and -- once established -- can be a very productive pasture.

Ischaemum aristatum (also known as I. indicum) (batiki blue grass)

A pasture grass of the South Pacific, batiki blue is used extensively in Fiji in rolling to hilly lands (Ranacou, 1972). Reynolds (1977d) reports that it is suited to acid soils above 600 m in Western Samoa. A creeping, highly stoloniferous grass, batiki is difficult to use in legume/grass mixtures because of its competitive nature, but it is very shade tolerant and does well under young coconuts.

Management. Batiki can stand heavy grazing, is about equal to signal grass in shade tolerance, grows best in pure stands, and so far no legume has been found that can grow in mixtures with it. Dr. George Osborne (pers. comm., 1972) says that batiki blue grass pastures on the island of Malaita are the best commercial pastures in the Solomon Islands. It has a high nitrogen requirement and may compete severely with coconuts for that reason (Ranacou, pers. comm., 1972). In Fiji, batiki is a poor producer during winter (which is also the dry sea-son), and for that reason koronivia grass is more favored.

Because of its low, creeping habit, nut collec-tion is easy in batiki pastures. The grass is widespread in Fiji, Solomon Islands, and Western Samoa.

Special Problems or Strengths. Relatively easy
and inexpensive to establish from cuttings, batiki
can tolerate poor management, but may not tolerate
dry conditions as well as palisade grass (Reynolds,
1976b). Its poor winter growth (due to lower temperatures plus dry conditions in Fiji?) -- particularly
at high stocking rates -- is a drawback.

Digitaria decumbens (pangola grass)

Pangola is only moderately suited for use under
coconuts (Figure 7.7). It really isn't shade tolerant enough for young or closely-spaced palms, but
can be used to advantage in groves where palms are
spaced widely (e.g. 75 to 100 palms per ha) or where
palms are old and the canopy is thin.

Pangola is easily established, high-yielding,
adapted to many soils, and can withstand very heavy
grazing.

Management. Pangola grows well on coastal
sands, heavy clays, and somewhat waterlogged areas.
It does not prosper in heavy shade, where it assumes
upright growth and can be grazed out. In an experiment in Jamaica (Coconut Industry Board, 1962-71)
pangola was established in 14-year-old palms spaced
10.6 m apart (= 88 palms/ha). The pastures and palms
were fertilized with 80 kg of ammonium sulfate after
every grazing (4 to 5 grazings per year), except for
two grazings about 6 months apart when the coconuts
were fertilized with 4.4 kg each of a complete mixed fertilizer, applied in a wide circle around the
palms. Two years after the system was begun, palm
yields had increased -- from 14.7 nuts/palm to 21.8
in the fertilized control (natural pasture), and
from 11.9 to 41 nuts/palm in the fertilized pangola
pasture. Cow days per ha per year for the natural
and pangola pastures averaged 270 and 720 days,
respectively (Coconut Industry Board, 1964). In a
later experiment (Coconut Industry Board, 1967), an
ungrazed pangola pasture caused a 9 per cent reduction in nut yields as compared to the natural pasture.

Productivity of the Pasture. Pangola pastures
can be very productive. In the open in the humid
tropics, liveweight gains can reach or exceed 500 kg

Figure 7.7. A pangola grass (<u>Digitaria</u> <u>decumbens</u>)
pasture in 5-year-old palms. Note cattle damage
on lower fronds at upper right.

per ha. Production in the shaded coconut groves would be lower, but it is not known just how low it might be. In Jamaica, in a 15 month period pangola produced 5.1 grazing cycles, 316 cow days per ha, a carrying capacity of 0.86 cows per ha, 48.6 t/ha of fresh forage, and 10.5 t/ha of dry matter (Coconut Industry Board, 1971). In this same experiment, guinea grass was less competitive with palms than was pangola; nut yields under guinea averaged 83 per palm, while those under pangola averaged 75 nuts per palm, and natural pastures averaged 73 nuts per palm.

Compatibility with Legumes. Pangola mixes fairly well with centro, greenleaf desmodium, and Desmodium canum.

Special Strengths and Problems. Pangola is high yielding, fairly palatable, grows on soils of low to moderate fertility, and responds well to fertilization. It slows in growth during winter and may become unpalatable to stock if over-mature.

Stenotaphrum secondatum (buffalo grass, St. Augustine grass)

Buffalo grass has been used in the New Hebrides for many years. This creeping grass can be grazed closely and is well suited to dry, sandy sites. In a trial in the New Hebrides a buffalo/siratro pasture in a very thin stand of palms (41 per ha) carried 2.5 heifers per ha, and produced an average liveweight gain of 430 g/head/day (157 kg/yr), and 393 kg liveweight gain/ha/yr. At this stocking rate the siratro almost disappeared from the pasture. The trial was continued the next year at a stocking rate of 1.25 head/ha; the pasture yielded 610kg liveweight gain/head/day, which projected to an annual basis would produce 222 kg liveweight gain/head and 278 kg liveweight gain/ha/yr.

Brachiaria ruziziensis (ruzi grass)

Ruzi is a creeping grass that has shown promise under coconuts in Papua New Guinea and the New Hebrides (Hill, 1969; Weightman, 1977). Best adapted to fertile, well-drained soils, ruzi will require heavy fertilization on less fertile sites. It is very digestible, is very shade tolerant (Watson, 1977), has high feeding value, and is well liked by cattle. However, it has a shorter growing season

185

than signal grass. It may fit into some coconut/
pasture systems, but should be evaluated further.

OTHER FODDER CROPS

Leucaena leucocephala (leucaena, ipil ipil, koa haole)

Leucaena is an excellent pasture plant for use
on coconut farms. It can be planted as a hedge from
which the young branches can be cut and fed as cut-
and-carry fodder. It regrows rapidly after cutting
and can be harvested on a 60 to 90 day cycle.

Leucaena can be grazed successfully under a
browse system, but its tall growth could cause problems in nut collection. If grazed under coconuts,
it should be planted in 2 or 3 widely-spaced rows 2
to 3 m apart in the interrows of palms, and underplanted with a creeping grass such as cori, signal,
palisade, or koronivia. Such possible mixtures need
evaluation. In the open, leucaena mixes well with
guinea.

Leucaena can tolerate very high calcium soils
(also high pH); therefore it may fit into systems on
coral sand soils.

Leucaena is shade tolerant and high-yielding;
e.g. it has produced 15 to 20 t/ha of dry forage and
1,600 to 3,000 or more kg/ha of crude protein per/yr.

Leucaena could be easily combined with special
napier or guinea grass pastures in cut-and-carry
systems to provide high yields of nutritious forage.
It has been used to good advantage for dairy cattle
in Hawaii (Henke and Morita, 1954; Takahashi and
Ripperton, 1949; Kinch and Ripperton, 1962).

Sugarcane can be a very good source of feed for
cattle, particularly when the green tops are used as
fodder (Preston and Leng, 1978). Sugarcane tops and
leucaena can be fed together with very good results.

Pure Legume Pastures

Shading by palms can lead to situations where
shade-tolerant legumes can survive better under
grazing than less shade-tolerant grasses. In heavily shaded situations the pastures may become domi-

nated by legumes. In many situations shading may cause the usually-dominant grasses to become more on a par competitively with legumes, and -- provided severe overgrazing doesn't occur -- a balanced legume/grass pasture may result. Legume-dominant or nearly pure legume pastures are not as serious a problem in the tropics as they would be in the temperate zones, because most tropical legumes do not seem to cause bloat.

On Niue island in the South Pacific, pure siratro pastures are grazed with good results (Lucas, 1972). In many cases where puero has been used as a cover crop, pure puero is grazed successfully without injury to the cattle.

Specialized pastures may be possible under coconuts, in situations where heavy shading and grazing cause pastures to become of pure legume composition, or legume dominant. In younger palms or in closely-spaced groves this situation may not be possible to overcome without thinning of palms, a step most farmers do not like to take, particularly in a well established, heavy-producing grove. One solution might be to grow pure grass pastures on more open lands, and -- if fencing arrangements permit -- allow the animals to graze on both the legume-dominant and grass-dominant pastures.

8
Improved Pasture Species

Not all tropical grasses or legumes are suitable for use under coconuts. However, a number of
plants that are or could be useful as cover crops, as
fodder or soilage crops, or as pasture plants will
be discussed because of their association with
coconut. A number of references were used in preparing this chapter. These include the following:
Imperial Agricultural Bureau, Great Britain (1944),
Whyte et al. (1959,1953), Norris (1967, 1969),
Appadurai (1968), Purseglove (1968), Sproat (1968),
Santhirasegaram, et al. (1969), Moore (1970), Semple
(1970), Jones (1972), Hugh (1972b), Manidool (1972,
1974), Mc Ilroy (1972), Plucknett (1972a), Ranacou
(1972b), Republic of the Philippines (1972),
de Guzman (1974, 1975), Humphreys (1974), Javier
(1974), Agency for International Development (1975a,
1975b), de Guzman and Allo (1975), Javier (1976),
Philippine Council for Agriculture and Resources
Research (1976c), Reynolds (1976b), Ericksen (1977).

DESIRABLE CHARACTERISTICS FOR FORAGE SPECIES FOR USE UNDER COCONUTS

In order to obtain good pasture growth under
coconuts, pasture species should have some of the
following characteristics (Plucknett, 1972a, 1972b):
  (1) be shade tolerant (to some degree).
  (2) should be capable of being grazed to within
      8 to 10 cm from the ground, so nuts can be
      collected easily.
  (3) be capable of enduring trampling by large
      cattle numbers.
  (4) should be perennial, probably creeping; if
      not, should have good seed production and a

188

vigorous seedling stage.

(5) should be palatable, and acceptable to animals.

(6) should have reasonably uncomplicated nutritional requirements; should not compete seriously with coconuts for nutrients.

LEGUMES

Because cover crops are probably the most widely used legumes in coconuts, the major species will be discussed first, followed by forage and pasture legumes. For a discussion of the use of cover crops in coconut, see Chapter 3.

MAJOR COVER CROPS

Pueraria phaseoloides (puero; also known as tropical kudzu)

A creeping, twining, pioneering, perennial legume from Malaysia, puero is the most important cover crop for tropical plantation crops such as rubber, oil palm, cacao, and coconut. It is exceptionally vigorous, competing with and smothering weeds.

Description. A climbing or creeping perennial herbaceous legume (Figures 8.1a,b), spreading by long vigorous runners and rooting at nodes to provide a deep, smothering mat: leaves with 3 leaflets, leaflets large, round and hairy; inflorescence a raceme, borne in leaf axils; flowers large, varying in color from blue and white, pink, to pale bluish purple; seed pods cylindrical, sometimes somewhat flattened, about 8 to 13 cm in length, 0.6 cm in diameter, pods mature irregularly; seeds small, about 3 mm long, oblong.

Propagation and establishment. Can be established from stem cuttings or seed. When vegetatively propagated in planting circles on a 3 X 3 m (or less) grid, it will spread rapidly to cover the ground and smother weeds. Seeding rate is about 5 kg per ha alone or 1 to 3 kg per ha in mixtures with other legumes or grasses. There are about 80,000 seeds per kg. Belongs to the cowpea group, so usually will not require inoculation. Seedling growth is rapid.

Climatic adaptation. Widely adapted to tropical areas with medium to high rainfall and moderate
to high temperatures: cannot stand cold temperatures.
Can survive long dry seasons because of a deep root
system.

Soil adaptation. Grows on wide range, including acid soils. Pioneers on newly cleared land.
Responds to P fertilizers. Can withstand some waterlogging.

Management. Most useful as a cover crop; when
grown alone will form a dense mat which can compete
with coconuts unless controlled by grazing or cutting. Cannot tolerate, and will not persist under,
heavy grazing. Can withstand light to moderate
shading. Does not require inoculation with Rhizobium.
Can provide early grazing but will not persist.
Should be sown with more persistent legumes such as
centro.

Calopogonium mucunoides (calopo).

A short-lived perennial from South America,
calopo is widely grown in the tropics as a cover
crop in plantation crops and as a pasture legume.
It is hardy and naturalizes readily.

Description. A trailing, climbing, or creeping
short-lived perennial herbaceous legume (Figure 8.2);
leaves with 3 leaflets, leaflets ovate, large 3 to
10 cm long, 2 to 7 cm wide; inflorescence borne in
leaf axils, composed of 2 to 4 flower groups; flowers
blue, small, 7 to 10 mm long; pods brown, narrow, 3
to 4 cm long, 0.5 cm wide, covered with fine brown
hairs, bearing 2 to 4 seeds; seeds flattened, square,
light brown.

Propagation and establishment. Reproduces by
seed, becomes easily naturalized. Seeding rate in
pure stands is 5 to 8 kg per ha. There are about
73,000 seeds per kg. Slower to establish than puero.
Produces dense mat 0.3 to 0.6 m high in 5 months.
Belongs to the cowpea inoculation group, so usually
will not require inoculation.

Climatic adaptation. Adapted to warm tropical
areas with rainfall of 1,000 to 1,400 mm or more.
Cannot stand cold temperatures.

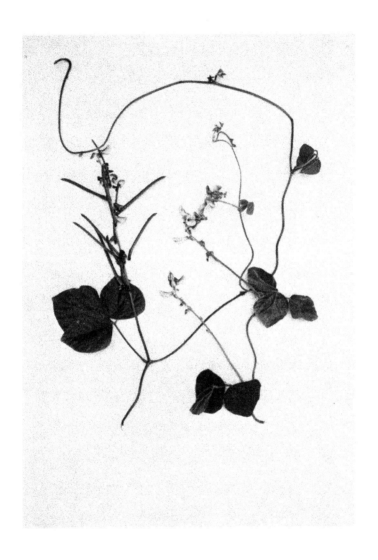

Figure 8.1a. Tropical kudzu (<u>Pueraria</u>
<u>phaseoloides</u>). Leaves, flowers, pods.

191

Figure 8.1b. Tropical kudzu (<u>Pueraria</u>
<u>phaseoloides</u>).

CALOPOGONIUM MUCUNOIDES
Common Name - Calopo

Figure 8.2. Calopo (<u>Calopogonium</u> <u>mucunoides</u>).
Plant habit, pods, seed. (illustration courtesy
Dr. Emil Javier, University of the Philippines
at Los Banos).

193

Soil adaptation. A pioneer legume, grows well on newly cleared lands.

Management. Can stand some shading. Valued as a cover crop. Not very palatable for livestock. Can be grown alone or in mixtures with leguminous cover crops such puero and centro or with tall tropical grasses.

Centrosema pubescens (centro).

A trailing or twining, climbing perennial legume from tropical America used widely as a cover crop and as a pasture plant. It is hardy and drought resistant.

Description. A twining or trailing, climbing perennial herbaceous legume (Figure 8.3); stems long, often rooting at the nodes; leaves with 3 leaflets, leaflets ovate with pointed tips, hairy, 5 to 12 cm long and 3 to 10 cm wide; inflorescence a raceme; flowers mauve, pea type; pods dark brown, 12 or so cm long, narrow, with a sharp tip; bearing up to 20 seeds; seeds large, 5 x 4 mm, dark brown or brownish black.

Propagation and establishment. Reproduces by seed but does root at nodes. Requires specific inoculant ("centro"). Seedling growth can be slow under shade, and in general is slower than calopo or puero. May be sown on freshly burned lands, broadcast and harrowed in, or drill sown. Seeding rate is 1 to 6 kg per ha. There are about 40,000 seeds per kg. Hard seed may be a problem; pour boiling water over the seed and allow to stand for 30 minutes, then wash and dry. Treat for hard seed just before sowing.

Climatic adaptation. Used widely in the warm tropics with rainfall above 1,000 mm. Does not grow well in cooler areas, but can survive frost.

Soil adaptation. Fairly well adapted to a wide range of soils, including acid soils of medium fertility. Responds to P. Can tolerate poorly drained conditions and can be used on alluvial soils.

Management. Can stand shading, especially when mature. Valued as a cover crop, especially in mixtures with the more rapid-growing puero and calopo,

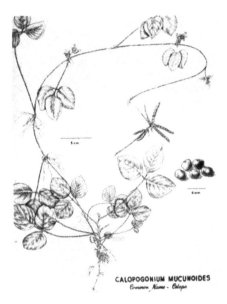

CALOPOGONIUM MUCUNOIDES
Common Name - Calapo

Figure 8.3a. Centro (Centrosema pubescens).
Plant habit, pods, seeds.

Figure 8.3b. Centro (Centrosema pubescens).
Close-up of centro in the field.

but becomes dominant when mature. Is highly persistant and can withstand a prolonged dry season. As a cover crop can compete with the crop and will climb on anything.

Can withstand heavy grazing. Suited to grass/ legume mixtures with guinea, napier, pangola and para grasses. Especially tolerant of shading by tall tropical grasses.

Very susceptible to 2,4-D or 2,4,5-T or other hormone sprays. To control weeds in centro pastures, use directed spraying techniques.

FORAGE AND PASTURE LEGUMES

Desmodium canum (kaimi clover).

A small prostrate to upright forb or shrub (Figure 8.4) which is a native of the Caribbean Islands, kaimi clover is an exceptionally hardy and persistent pasture legume (Younge et al., 1964). It has become naturalized in many Pacific islands.

Description. A small (0.2 to 0.5 m tall), upright, somewhat herbaceous to shrubby perennial legume; stems of two types, prostrate and upright, each of which bears a distinctive type of leaf; leaves with 3 leaflets, the leaves on upright stems are lanceolate with a white marking in the center on either side of the midrib, leaves on prostrate or creeping stems are oval to round and without leaf markings; inflorescence a raceme; flowers reddish or lavender; pods 2 to 4 cm long, segmented, hairy, and adhering readily to fur or clothing, bearing 4 to 7 seeds; seeds light brown, kidney-shaped, variable in size.

Propagation and establishment. Established mainly from seeds. Spreads naturally because the hairy pod segments adhere easily to many surfaces and are spread widely by man and animals. Sowing rate is 5 to 10 kg per ha. There are about 170,000 to 500,000 seeds per kg (Rotar and Urata, 1966). Can be drill or broadcast sown. Would probably benefit from inoculation; use specific Desmodium inoculant.

Climatic adaptation. Widely adapted in the more humid tropics (1,000 to 2,500 mm or more).

196

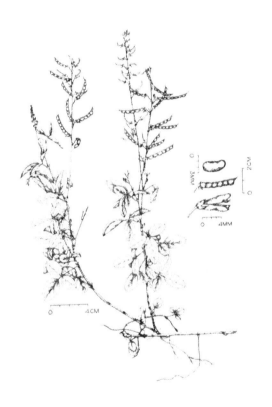

Figure 8.4.   Kaimi clover (<u>Desmodium</u> <u>canum</u>);
plant habit, flower, pod, seed.

197

Soil adaptation. Widely adapted, kaimi has a very low P requirement and can withstand very high levels of Al and Mn. It grows readily in soils where stylo, well known for its adaptation to infertile soils, also grows well.

Management. Kaimi is easily managed, largely because it is so hardy and persistent. It can tolerate shading, can persist under heavy or abusive grazing, and can survive in pastures where nitrogen fertilizers are used. It probably fixes only a small amount of nitrogen and because of its persistence is always present to provide protein for the grazing animal. Moderate to heavy grazing favors this small, low-growing legume.

Kaimi mixes well with stoloniferous or low creeping grasses like pangola. It would appear to be an excellent candidate for trial under coconuts where levels of management are likely to remain low.

Desmodium intortum (green-leaf desmodium)

A trailing perennial legume which is a native of Central and South America, green-leaf desmodium is used widely as a pasture legume in the wetter subtropics.

Description. A trailing, creeping perennial herbaceous legume (Figure 8.5); spreading by long vigorous pubescent stems which root at nodes to provide a deep vigorous mat; leaves with three leaflets, leaflets ovate, upper surface dark green with very fine gray hairs and with characteristic reddish brown to purple flecking, lower surface grayish-green with many white silky hairs; inflorescence a raceme; flowers lilac to pink, giving rise to pods; pods segmental, 8 to 12 segments, hairy, each segment bearing a single seed; seeds light brown, kidney-shaped, 1.5 mm long, 1 mm wide.

Propagation and establishment. Progagation is generally by seed (1-2 kg/ha) in an adequately prepared seed-bed. There are about 750,000 seeds per kg. It has established well after broadcasting into a cool ash seed-bed in periods of reliable rainfall. Where seed is not available stem cuttings can be used. Inoculation of seed is necessary (specific desmodium inoculant) before planting.

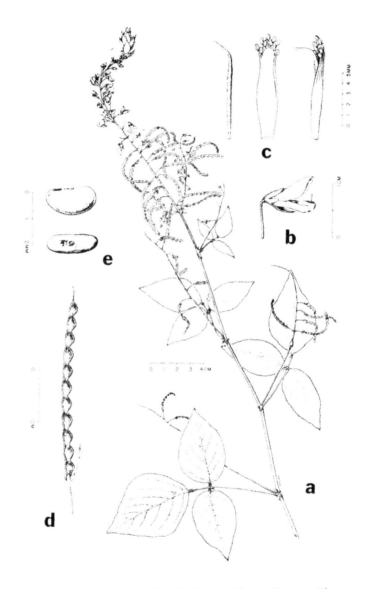

Figure 8.5. Greenleaf desmodium (Desmodium
intortum). (a) Plant habit, (b) flower, (c)
floral parts, (d) pod containing seed, and (e)
seed. (Illustration from Younge, Plucknett and
Rotar, 1964).

Climatic adaptation. Greenleaf requires an annual rainfall of at least 1,000 mm and will withstand some waterlogging. It is not particularly drought tolerant and grows best under warm subtropical conditions.

Soil adaptation. It is adapted to a wide range of soil conditions from sandy-textured, light soils to loams and heavy clays. It also tolerates acid conditions and is adapted to some poorly drained situations. It is most responsive to P and K and is capable of fixing large amounts of nitrogen.

Management. Once established, green leaf desmodium spreads vigorously and can smother many weeds under adequate management. More lenient grazing is needed during the establishment phase if the legume is to provide good cover. Once mature woody runners have developed, this legume has considerable tolerance to heavy grazing pressure.

Glycine wightii (formerly G. javanica), glycine, perennial soybean

A trailing perennial legume indigenous to Africa, glycine is found widely throughout the East Indies and tropical Asia. Once established, it is a vigorous twining plant which will root at the nodes when in contact with the soil.

Description. A twining, trailing perennial herbaceous legume (Figure 8.6); stems long, rooting at nodes; leaves with three leaflets, leaflets broad with fine short hairs on the undersides; inflorescence a raceme; flowers small, white; pods 2 to 5 cm long, dark brown-black with light brown hairs; seeds small.

Propagation and establishment. Established by seed (2-4 kg/ha). There are about 154,000 seeds per kg. Seedling vigor is inferior to most other legumes, but once nodulation becomes effective, and under favorable conditions of soil fertility, growth and production improve markedly. Adequate seed-bed preparation and inoculation are essential. Requires cowpea inculant.

Climatic adaptation. Glycine requires an annual precipitation of 750 to 1,750 mm, but can tolerate seasonal droughts due to its deep tap-root and strong lateral root system. Glycine is late flowering and thus has a long season of vegetative growth. It is more adapted to the cooler sub-tropics than most of the other tropical legumes.

Soil adaptation. Compared to other tropical legumes, glycine is more specific and demanding in its soil requirements. It is intolerant to water logging, being adapted to deep, well-drained soils. It has a high demand for P and K. Glycine also seems to be more sensitive to excessive levels of soil Mn and Al, and its response to liming may result from the effect of lime on the availability of Al, Mn, and P.

Management. Because of slow seedling establishment, early management of glycine in a grass-legume stand is critical if the legume is to persist. Once established, glycine persists well under grazing and exhibits considerable shade tolerance under plantation crops. Because of its climbing nature it could smother young plantation crops if clearing or trimming are not practiced.

Figure 8.6. Glycine (Glycine wightii, formerly G. javanica) in the field.

Leucaena leucocephala (leucaena)

Leucaena is a perennial leguminous shrub native to Central and South America (Takahashi and Ripperton, 1949); Kinch and Ripperton, 1962; Gray, 1968; Oakes, 1968). It is used as a shade plant in smaller plantation crops, as a green manure, or as a high protein forage.

Description. An upright, small (2 m) to large (20 m or more) perennial shrub or tree (Figure 8.7); taproot deep; leaves large, bipinnate, 15 to 20 cm long, individual pinnae 10 cm long; leaflets small, rather oblong, borne in pairs along the pinnae; flowers white or yellow, borne in ball-like fluffy clusters, giving rise to pods; pods brown, flat, 20 cm or so long, bearing many seeds; seeds long-oval, flattened, 6 mm long.
Leucaena has been grouped into 3 distinct types: the short, bushy Hawaiian type; the tall, sparsely-branched El Salvador type; and the tall, strongly-branched Peru type.

Propagation and establishment. Leucaena can be established either by direct seeding, stem cuttings, or seedling transplants. Direct seeding results in slow seedling emergence, and adequate mechanical or chemical weed control must be practiced to reduce competition from other herbaceous or grass species during establishment. Leucaena is generally estab-lished in rows 2 to 3 m apart to allow for interrow mechanical weed control and for maintenance of com-panion grass forage species. In smaller holdings, leucaena might best be propagated in hedgerows, and hand cutting and feeding practiced. Before planting, seed requires immersion in hot water (80 C) for 3 minutes (Gray, 1962) to ensure good germination. Inoculation with specific "leucaena" rhizobium is essential, and lime pelleting has been found to be advantageous in acid soils.

Climatic adaptation. Leucaena is adapted to tropical areas with an annual rainfall in excess of 760 mm. However, its deep rooting habit allows this plant to exhibit considerable drought tolerance. During very dry periods, the plant may shed its leaves.

Soil adaptation. One of the few tropical legumes adapted to calcareous soils, it responds to both lime and P on acid soils.

Management. Cattle will readily graze young leaves and twigs. During the establishment phase, plants should be grazed when 0.6 to 0.9 m in height and then allowed a period of recovery. In later years, heavier grazing will be required at times to keep the shrub under control. Lopping and slashing may be necessary if plants become too woody. The toxic alkaloid (mimosine) present in young leaves and shoots of leucaena causes loss of hair in non-ruminants and when making up a high proportion of the diet in ruminants, has been reported to cause reproductive problems and enlarged throid glands (CSIRO, 1975). However, as a general rule when used in moderation with a combination of grass forage, the advantages of this high yielding, high protein feed are too great to be overlooked.

Leucaena mixes especially well with guinea grass, either when planted in spaced rows with guinea grass between the rows, or in open mixtures. Any upright or stoloniferous grass with some shade tolerance should do well in combination with this legume. Cattle browse the leaves, pods, and young twigs readily.

For coconut pastures, leucaena could be planted in spaced rows in combination with guinea or other compatible grasses, or in fencerows or hedgerows where the forage could be cut and fed as fresh-cut forage for the animals.

Macroptilium atropurpureum (Siratro).

The result of a cross between two Mexican strains of Phaseolus atropurpureus, siratro is now widely adapted to many areas where improved pastures are being developed in the tropics and sub-tropics. Dr. E. M. Hutton of CSIRO in Australia bred and developed siratro.

Description. A strongly stoloniferous, herb-aceous perennial legume (Figure 8.8); stems (stolons) creeping, long, rooting at the nodes; leaves with three leaflets, leaflets broad, characteristically lobed, green on the upper surface and covered with silvery-gray fine hairs on the lower surface; in-florescence a raceme, 10 to 30 cm in length; flowers deep red or purple, giving rise to pods; pods nar-row, straight cylindrical, about 8 cm long, carrying about 12 to 13 seeds, shattering at maturity; seeds large, ovoid, dark brown.

203

Propagation and establishment. Siratro establishes best in a well prepared seed bed, but can also
establish after broadcasting into ash or with minimum soil disturbance. Inoculation is desirable (belongs to cowpea group) but it can nodulate freely
with native rhizobium strains. Its ability to seed
freely aids its spread by seedling regeneration.
Seeding rate is 1 to 3 kg per ha. There are about
80,000 seeds per kg.

Climatic adaptation. Being sensitive to low
temperature, siratro, as with most of the other
"tropical" legumes, is best adapted to moist, warm
subtropical climates. Its best performance is obtained in areas receiving from 750 to 1,750 mm of
annual rainfall. Because of its large, deep taproot,
siratro is very drought tolerant and can grow under

Figure 8.7. Leucaena (Leucaena leucocephala)
(formerly known as L. glauca). Portion of
branch showing leaves, pods, puffball-like
flowers and seeds.

lower rainfall conditions. However, in wetter areas (above 1,750 mm) of the tropics the use of siratro is limited by its susceptibility to the disease, Rhizoctonia solani.

Soil adaptation. Siratro is versatile in its soil requirements, especially where drainage is adequate, from light-textured sandy soils to heavy clays. It responds to improved soil fertility and is intermediate to glycine and stylo regarding its response to P fertilization.

Management. As with most of the other legumes, it combines well with many of the improved tropical grasses. Under moderate shading by plantation crops shows good evidence of persistence and spread.

Stylosanthes guyanensis (stylo)

A shrubby perennial legume from Brazil, stylo is exceptionally hardy and adapted to infertile soils. It is used for pasture, but is not shade tolerant, so should only be used in older, widely-spaced coconuts.

Description. An erect, sub-woody perennial shrub (Figure 8.9), growing up to 1.6 m or so in height; leaves with 3 leaflets, leaflets long, narrow and pointed; flowers small, yellow; pods small, producing single seeds; seeds yellow, small.

Propagation and establishment. Propagated by seed. Sowing rate is 2 to 5 kg per ha. There are about 350,000 seeds per kg. Can be sown on freshly-burned lands, broadcast and harrowed in, or drill sown about 0.5 cm deep. Does not require inoculation (belongs to cowpea group) but in some cases may benefit from inoculation. One cultivar, "Oxley Fine Stem," requires a specific inoculum, "fine leaf stylo," but this variety is not likely to be used under coconuts.

Climatic adaptation. Well adapted to the warm tropical areas. Cannot tolerate cold temperatures, and may be killed by frost. Can grow well under high rainfall, 3,500 mm or more, but can be grown under rainfall as low as 900 mm. Can withstand prolonged dry weather after mature root system develops.

PHASEOLUS ATROPURPUREUS DC.

Common Names — Purple bean, Siratro

Figure 8.8. Siratro (<u>Macroptilium atropurpureum</u>).
Above - closeup view of siratro in the field.
Below - plant habit, seeds, pods. (Illustration
courtesy of Dr. Emil Javier, University of the
Philippines at Los Banos).

Soil adaptation. A major strength is its ability to grow under very low fertility conditions. Can tolerate high levels of active Al and low levels of P in the soil. Responds to P fertilizers. Grows well on coastal sands and shallow, rocky soils.

Management. Should not be planted under young or closely spaced coconuts because of poor shade tolerance. There are better legumes for most situations, except for infertile soils carrying older, widely-spaced palms. Moderately palatable for stock, especially at later growth stages when leaves are dry; can then be "hayed off." Cannot stand fire and stands may be thinned, if fire results. Can stand moderately heavy grazing. Very tolerant of 2,4-D and other hormone sprays.

Figure 8.9. Stylo (Stylosanthes guyanensis). Close-up view of a stem showing leaves and flowers.

GRASSES

A number of grasses have been used under coconuts; others would appear to have promise. Certain grasses may be chosen for use under coconuts because of a number of reasons, including availability of seed or propagating material, previous experience of farmers or extension workers, or the soil and rainfall situation encountered. Some grasses are more suitable for pasture; others, like napier, may be more suitable as green, cut fodder.

Some of the grasses discussed here are also discussed in Chapters 5 (Natural Pastures) and 7 (Improved or Established Pastures).

Brachiaria brizantha (palisade grass, sometimes called signal grass)

A very desirable grass from Africa, palisade grass is an outstanding pasture plant under coconuts (Anonymous, 1962). It is shade tolerant, competes readily with weeds, and is drought resistant.

Description. A loosely tufted, coarse perennial grass (Figure 8.10); spreading by rhizomes; culms erect, ascending, 40 to 140 cm high; nodes hairy or sometimes smooth; leaf sheath more or less softly hairy; ligule very short, with a fringe of hairs about 1.5 mm long; leaf blade 5 to 45 cm long, 5 to 12 mm wide, mostly hairy on the surfaces, with margins which are rough to the touch; inflorescence a panicle of 2 to 7 drooping to horizontal hairy spike-like racemes which arise from one side of the purplish panicle axis; racemes 5 to 15 cm long, each with a hairy axis; spikelets swollen, elliptic, large, about 5 mm long, often purple, arranged in a single row on the rachis, conspicuous purple stigmas.

Propagation and establishment. Propagated vegetatively from stem cuttings or rhizomes which can be harvested from mature pastures or special nursery areas. Can be planted in furrows, in planting holes, or in clean-weeded circles about 1 m or so in diameter. In Western Samoa, WSTEC plants palisade grass in such cleared circles spaced in a 3 X 3 m or 4 X 4 m grid; a 3 X 3 m grid pattern takes about one year to cover the ground, a 4 X 4 or 5 X 5 m grid takes about 18 months to cover. Such a system is very inexpensive to establish. In this system a small bundle of stem cuttings taken from special

nurseries are planted in each planting circle.
Highly competitive with weeds, it takes over and
covers the ground rapidly.

Climatic adaptation. Best adapted to the warm
moist tropics where rainfall exceeds 900 to 1,000 mm.
Deep rooted, it can withstand some drought and can
produce forage during dry periods.

Soil adaptation. Adapted to a wide range of
soils, especially moist lowlands.

Management. A valuable pasture grass. Can
withstand fairly heavy grazing. Produces good
quality palatable forage and is very useful for
specialized uses, such as pasture for young stock.
Mixes well with legumes such as phasey bean
(Macroptilium lathyroides).
Very shade tolerant and well suited to coconut/
pasture systems.

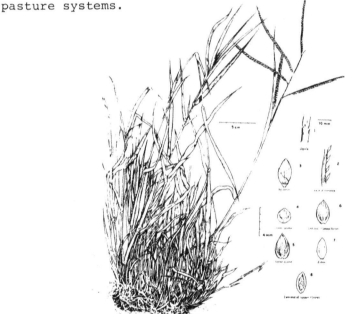

Figure 8.10. Palisade grass (Brachiaria
brizantha). Plant habit; 1) ligule; 2) axis of
raceme; 3) spikelet; 4) lower glume; 5) upper
glume; 6) lemna of lower floret; 7) palea; 8)
lemna of upper floret. (Illustration courtesy
of Dr. Emil Javier, Univ. of the Philippines
at Los Banos).

Brachiaria decumbens (signal grass)

A native of Uganda, signal grass is a strongly stoloniferous pasture grass which is increasingly finding favor in the wet tropics.

Description. A low-growing, creeping perennial grass (Figure 8.11); spreading by stolons which root at the nodes; culms prostrate to ascending, up to 30 to 60 cm or so high; ligule conspicuous; leaf blades 18 cm or so long, bright green; inflorescence a lax panicle with 2 to 6 short spike-like racemes; racemes borne at right angles to the panicle axis (hence its name "signal" grass), with 1 or 2 rows of spikelets; spikelets large, hairy, crowded on one side of a broad, flattened, and winged rachis. B. decumbens is difficult to distinguish from B. brizantha.

Propagation and establishment. Propagated vegetatively and by seed, following discovery in Australia that germination could be enhanced by storage of seed for 10 months or by treatment with concentrated sulfuric acid. A sowing rate of 4 to 6 kg per ha is suggested; there are about 700,000 seeds per kg.

For vegetative propagation, stolons can be cut and planted in furrows or in planting holes. A vigorous grass, it should cover the ground quickly. Graze early to encourage spreading and vigorous stooling, and to ensure adequate competition with weeds when heavy use is expected.

Soil adaptation. Grows well on a wide range of soils, but does best under well-drained and fertile conditions.

Management. One of the most productive of the perennial wet tropical grasses, both from the standpoint of forage yields as well as animal production; can be intensively managed to obtain good production. When nitrogen fertilizers are applied to signal grass prior to cool periods, it will produce forage when other grasses are producing little or nothing. Used in this way, it offers opportunity to obtain year-round feed production by use of specialized pastures and strategically-timed nitrogen fertilizers.

Signal grass should be heavily stocked and grazed in order to keep the sward low, young and palatable.

Mixes well with centro and probably with other creeping perennial tropical legumes.

Figure 8.11. Signal grass (<u>Brachiaria</u> <u>decumbens</u>).

<u>Brachiaria humidicola</u> (koronivia grass)

Koronivia is highly regarded for coconut/pasture systems in Fiji, especially because it grows quite well during dry cool winter months when it can carry as many as 2.5 animals per ha (Ranacou, 1972b). It is recommended with siratro on coral sandy soils and with centro on rolling to hilly lands. It appears well suited for dry, better-drained soils. On coral sands it grows vigorously and forms a dense cover.

Koronivia can be established vegetatively by stem cuttings spaced 1 m apart in rows between the coconuts. It produces a small amount of viable seed and is very shade tolerant. Nut collection is easy in koronivia pastures because it does not grow tall.

<u>Brachiaria miliiformis</u> (cori grass)

A creeping perennial grass from Asia, cori grass is probably the most promising pasture grass under coconuts. Exceptionally shade tolerant, it establishes easily, grows rapidly, is very palatable to animals, and is very productive, while at the same time being less competitive with coconuts than other

211

grasses.

Description. A creeping perennial grass (Figure
8.12); stems somewhat slender, up to 1 m long, ascending, 10 to 30 cm high, rooting at hairless nodes;
leaf sheaths smooth except for fine hairs on the
nodes and margins; ligule a thickened ridge of stiff
white hairs; leaf blade 2 to 12 cm long, 5 to 8 mm
wide, tapering to a sharp tip, rounded at the base;
inflorescence a very open panicle with 2 to 4 widely
spaced racemes; racemes 1 to 5 cm long; spikelets
pale green, attached singly in 2 rows on one side of
the raceme, rachis 3-cornered, spikelet with short
stalks, elliptical, hairless; lower empty glume
short, wrapped round the base of the spikelet; upper
glume as long as the spikelet, 7-nerved, stigmas
purple.

Propagation and establishment. Can be established from stolon (stem) cuttings or seed. Grows
rapidly from vegetative cuttings and covers the
ground quickly.

Climatic adaptation. Widely adapted to tropical areas with medium and high rainfall and moderate
to high temperatures.

Soil adaptation. Apparently adapted to wide
range of soils. In Sri Lanka, Senaratne (1956)
stated that it was common throughout the island.

Management. Quite shade tolerant. Vigorous
growth, can withstand fairly heavy grazing. Competes
well with weeds. Can be grown in mixture with
legumes, although the full range of potential legumes
for use in mixtures has not been evaluated.

Brachiaria mutica (para grass)

Formerly known as Panicum purpurascens, para
grass is a native of Africa. A coarse, trailing
grass which roots at the nodes, para is especially
valuable because, although it is somewhat drought
resistant, it grows luxuriantly in swampy, nearly
waterlogged lands.

Description. A stout, creeping, spreading
perennial grass (Figure 8.13); stolons to 5 m or so
in length; culms decumbent to ascending, rooting at
the base, stoloniferous, nodes densely covered with

212

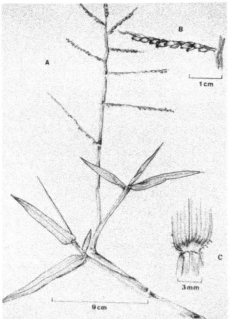

Figure 8.12. Cori grass (<u>Brachiaria miliiformis</u>).
Above - closeup view of cori grass in the field.
Below - plant habit, detail of ligule and
raceme.

213

Brachiaria mutica *(Forsk.) Stapf*
Common name Para grass

Figure 8.13. Para grass (Brachiaria mutica).
Plant habit, 1) spikelet; 2,3) glume; 4,5) lemna
and palea; 6) ligule and leaf sheath; 7) rachis.
(Illustration courtesy of Dr. Emil Javier,
University of the Philippines at Los Banos).

214

long soft hairs; sheaths softly hairy or smooth on
upper portion with densely hairy collar; leaves
10 to 30 cm long, 5 to 15 mm wide, flat, smooth;
inflorescence a panicle, 15 to 30 cm long, densely
flowered branches somewhat separated, 2.5 to 10 cm
long; spikelets 3 to 5 mm long, occasionally with
light purple tinge.

Propagation and establishment. Para rarely sets
seed and is usually propagated by stolon cuttings
which sprout and take root rapidly. Cuttings can be
planted in furrows or can be broadcast on the soil
surface and disced in.
In many places para grass will invade lands and
grow naturally. It is one of the grasses which can
be a component of natural pastures under coconuts.

Climatic adaptation. A plant of the warm, moist
tropics and sub-tropics, para grass is mainly a plant
of poorly drained or wet areas. It is adapted to
areas receiving from about 1,000 mm of rainfall to
2,500 mm or more. It is best adapted to tropical
lowlands and coastal areas.

Soil adaptation. Para grass responds readily
to fertilization and generally grows best in soils
with medium or better fertility. Its primary advantage is a remarkable ability to grow in waterlogged
or nearly saturated soils. It can grow in slightly
brackish water.

Management. Under high moisture conditions para
grass forms dense, vigorous pure stands. It can be
grazed, but cannot withstand close, continuous grazing. However, it can be stocked heavily, but it may
be best to graze it in rotation; occasionally it
should be mowed to reduce clumping and to remove
coarse stem growth. It is sometimes used as a cut-
and-carry or green-chop forage in Hawaii, particularly for dairy cattle.
Para is somewhat shade tolerant and therefore
will grow vigorously under coconuts. Legumes which
can be grown with para are: greenleaf desmodium,
phasey bean, and stylo in poorly drained lands, and
centro and puero in better drained sites.
Para is one grass which can be used as reserve
pasture for dry season periods; to do this, fence
out wet or swampy spots, plant to para (or allow
natural growth to develop) and reserve this pasture
for later use.

215

Brachiaria ruziziensis (ruzi grass, congo grass)

A leafy, creeping perennial from Africa, ruzi
grass has become a useful pasture plant in the warm,
moist tropics and sub-tropics. In most situations,
because it has a shorter growing season and a high
soil fertility requirement, it probably will not
become as important as its relatives, cori, palisade,
para, or signal grasses.

Description. A tufted creeping perennial grass
(Figure 8.14); culms leafy, hairy, to 1.5 m high;
leaves numerous, hairy, broad; inflorescence a lax
panicle with 3 to 8 relatively long racemes; racemes
with 1 or 2 rows of spikelets crowded on one side
of a broad, flattened, and winged rachis (such as in
B. decumbens).

Propagation and establishment. A free seeder,
ruzi grass has received attention because of its
seed production. Seed germination is enhanced by
storage for 12 months or treatment with concentrated
sulfuric acid. Easily established from seed. Sowing rate is about 4 to 6 kg per ha. Propagated also
by vegetative cuttings and rooting at the nodes.

Climatic adaptation. Adapted to the warm tropics with rainfall of 1,000 mm or more. Has a shorter growing season than signal grass (B. decumbens).

Soil adaptation. Grows on a range of soils,
but is best suited to well-drained, fertile sites.
On poor soils will require heavy fertilization.

Management. Feeding value of ruzi grass is
high and does not decline with age as much as
other tropical grasses. Cattle readily accept it.
Legumes which mix well with other creeping grasses
will probably perform satisfactorily with ruzi grass.
It is not known how ruzi grass might perform
under coconuts, but cori, palisade, and signal
grasses would probably be better suited for such use,
especially because of their longer growing period
and less demanding fertility requirements.

Chloris gayana (Rhodes grass).

A tufted perennial, native to southern and
eastern Africa, and spreading somewhat from stolons
which root at the nodes, Rhodes grass is a widely-
adapted, hardy grass for lower rainfall areas of the

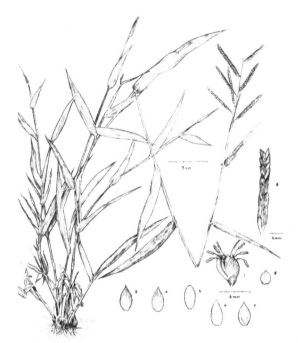

Figure 8.14. Ruzi grass (<u>Brachiaria</u> <u>ruziziensis</u>). Plant habit and inflorescence. 1) flower, 2) seed, 3-7) parts of the spikelet -- glumes, lemna, palea. (Illustration courtey of Dr. Emil Javier, University of the Philippines at Los Banos).

tropics and sub-tropics. For coconut lands, it may be of value for the lower rainfall areas where moderate to heavy grazing should be able to reduce the effects of its competition for moisture with the palms.

Description. A tufted, upright, stoloniferous perennial grass (Figure 8.15); roots vigorous, deep; stolons rooting at nodes; stems relatively fine, leafy, reaching as much as 120 to 150 cm in height, forming dense bunch-type habit of growth; inflorescence a group of 10 to 12 brownish-green seed spikes which radiate at upright angles from a common point; seeds very small.

There are four named varieties in Australia; Pioneer (common), Callide, Samford, and Katambora.

Propagation and establishment. Mainly established from seed, but can be vegetatively propagated.
Seeds are very small; there are from 3.3 to 4.4
million seeds per kg. Sowing rates are from 0.5 to
6 kg/ha.

If seeds are to be drill-sown, because of their
small size they must be placed at shallow depth, 0.6
to 1.2 cm. Sow before rains to ensure a good stand.

Climatic adaptation. Adapted widely in lower
rainfall (600 to 1,250 mm) tropics and sub-tropics.
Does not grow vigorously in high rainfall areas.

Soil adaptation. Widely adapted, but is especially noteworthy for its salt tolerance. Performs
best under good soil fertility conditions and will
not remain vigorous when fertility declines. Requires
well drained soils.

Management. Once established, Rhodes grass is
very hardy and can withstand heavy grazing. Should
be grazed frequently in order to prevent growth from
becoming rank, stemmy, and unpalatable.

Rhodes grass combines well with phasey bean,
siratro, and centro.

Cynodon plectostachyus (African star grass).

A creeping perennial native of East Africa,
African star grass has become popular in many parts
of the tropics. It has become a major pasture
species in Puerto Rico (Vicente - Chandler et al.
1974). Highly aggressive, it crowds out weeds and
other plants. It resembles bermuda grass (Cynodon
dactylon), having 3 to 20 racemes on the inflorescence. It can withstand heavy grazing and drought,
responds readily to fertilization, and is readily
accepted by cattle when young.

Dicanthium aristatum (angleton grass, alabang)

A decumbent, low-growing, creeping perennial
native of East Africa and India, angleton grass is
highly regarded as a pasture grass for coconut lands
in the Philippines. Although it is a prolific seeder, it is usually propagated vegetatively from stem
cuttings or rootstocks. Commercial seed is not
available.

Best adapted to areas with over 1,000 mm rainfall, angleton grass may have some salt tolerance.

218

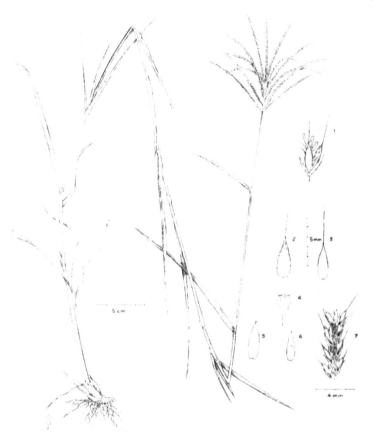

Figure 8.15. Rhodes grass (Chloris gayana).
Plant habit; 1)spikelet in flower; 2,3) spikelet,
two views; 4,5,6) glumes and lemma; 7) portion
of raceme showing spikelets. (Illustration
courtesy of Dr. Emil Javier, University of the
Philippines at Los Banos).

It can stand close grazing, is very palatable to
animals, and mixes well with legumes. Alabang is
the name of a cultivar developed by the Philippine
Bureau of Plant Industry (Javier, 1974).

Digitaria decumbens (pangola grass).

A low, creeping, perennial native of Southern
Africa, pangola grass has become a major pasture
grass in most parts of the warm moist tropics and
sub-tropics. It stops growing during cool winters
when minimum daily temperatures drop to about 11 C.

Description. A creeping to decumbent, mat-
forming perennial grass (Figure 8.16); stems long,
the creeping stems with conspicuously hairy nodes
that root readily to form new plants, decumbent stems
have smooth nodes; leaves 10 to 18 cm long, about 6
to 7 mm wide, smooth on both sides; ligule a prominent membrane; inflorescence a cluster of finger-
like branches, borne on long flowering stalks which
are produced on the decumbent stems; spikelets about
3 mm long; seeds seldom viable.

Propagation and establishment. Easily established by stem cuttings which are planted by hand in
furrows and then covered, or broadcast on the soil
surface and then disced into the soil. Once rooted,
the cuttings grow rapidly and cover the ground quickly. Rate of growth and spread is hastened by
phosphorus fertilizer.
Seeds are seldom viable.

Climatic adaptation. Essentially a plant for
the warm moist tropics, pangola does best in areas
receiving more than 1,000 mm of rainfall. It is not
well suited to areas with low rainfall. In areas
where minimum daily temperatures drop to 11 or 12 C,
it stops growing, and therefore it demonstrates a
marked seasonal growth.

Soil adaptation. Widely adapted to many soils,
from sands to heavy clays, pangola nonetheless responds readily to fertilizers, particularly nitrogen.
It can withstand some waterlogging.

Management. Under good nitrogen nutrition,
pangola can be highly productive. It is well accepted by cattle, especially when young, and can withstand heavy grazing pressure. Its sod-forming growth
habit makes it possible to grow several legumes in
combination with pangola; among there are centro,
greenleaf desmodium and kaimi clover.
Pangola should be grazed early when its digest-
ability and protein content are high; when mature
or when inflorescences are produced, the forage
quality drops considerably. Rotational grazing
helps to ensure production of good quality forage.
For coconuts, pangola may not be sufficiently
shade tolerant, especially for young or closely-
spaced palms. Pangola should be tested in older or
more widely-spaced palms to evaluate its suitability
for use in this system.

Figure 8.16. Pangola grass (<u>Digitaria</u> <u>decumbens</u>).
Plant habit, spikelet.

<u>Ischaemum aristatum</u> (batiki blue grass).

A native of tropical Asia, Batiki blue grass is
a creeping perennial with 2 racemes borne at the apex
of the flowering culms and with awned spikelets. It
is highly favored in Fiji as a pasture grass and is
common in pastures in Asia.

Description. A creeping, stoloniferous <u>peren-</u>
<u>nial</u> grass; <u>culms</u> decumbent to ascending, 15 to 150
cm high; <u>nodes</u> conspicuously bearded; <u>leaf sheath</u>
loose and hairy; <u>ligule</u> a membrane, hairy; <u>leaf</u>
<u>blades</u> narrow, shortly hairy, 5 to 30 cm long, 4 to
7 mm wide; <u>inflorescence</u> consists of a pair of spike-
like terminal racemes 4 to 7.5 cm long, with a joint-
ed, hairy rachis; <u>spikelets</u> paired, alike, one

221

sessile, the other stalked, with a 1 cm or so long
awn, spikelets falling at maturity with joints and
stalks attached.

Propagation and establishment. Mainly propagated vegetatively by stolon and culm cuttings.

Climatic adaptation. A plant of the warm wet
tropics. Does not grow well during dry cool winter
periods in Fiji.

Soil adaptation. Adapted in Fiji to wet rolling to hilly soils. Probably requires fairly high
soil fertility for best growth. Considered to be an
exceptionally heavy feeder for nitrogen, to the
extent that it may compete severely with coconut.

Management. Poor production during dry winters
in Fiji. Also used for pasture under coconuts in the
Solomon Islands. There is evidence in Fiji that
Batiki blue can depress nut yields of coconuts; this
was thought to be due to competition for nutrients
(Ranacou, 1972a, 1972b). Nut collection is easy
with Batiki blue.
Batiki blue grass competes well with weeds,
smothering and eliminating them from many swards,
especially on hill lands.

Melinis minutiflorus (molasses grass)

A native of Africa via Brazil, molasses grass
is a spreading, hairy perennial pioneer plant
(Hosaka and Ripperton, 1953). The soft leaves are
covered with sticky short hairs with a molasses-like
odor, hence its common name. Molasses grass seed is
very small and light, but viable, and for this reason
the plant is frequently used as a first-stage crop
in pasture improvement or development.

Description. A loosely straggling, tufted,
fast-growing perennial grass (Figure 8.17); stems
creeping, rooting at nodes, 2 m or so long, reaching
a height of 0.6 m or so; nodes long hairy; leaf
sheath densely hairy, the base of the hairs producing a sticky oil that smells like molasses; ligule
a fringe of white hairs 1 to 2 mm long; leaf blade
5 to 20 cm long, 4 to 12 mm wide, densely hairy,
sticky; inflorescence a dense, plume-like, many
branched panicle 10 to 25 cm long, spreading or compacted, purple in color; spikelets narrow, 2 mm long,

purplish, with awns 5 to 12 mm long.

Propagation and establishment. Propagated mainly by seed. Seeds tiny, light, about 13 to 15 mil-
lon per kg, tending to stick together. May be useful to use an inert material like sawdust to give a
more even distribution at planting. Sowing rate is
about 2 to 5 kg/ha.
Machine sowing of molasses grass is difficult
because of the light, sticky, small seeds. Broadcast sowing on poorly-prepared land can be successful; indeed, the ability of molasses grass to grow
and cover steep, rough or difficult lands is one of
its main virtues.

Climatic adaptation. Adapted to warm moist
tropics with rainfall of 1,000 mm or more. Once
established, it has some drought resistance.

Soil adaptation. Adapted to a wide range of
soils, provided that they are well drained. Can
grow well on soils of low fertility, but will respond
to N and P.

Figure 8.17. Molasses grass (Melinis
minutiflorus) in the field.

223

Management. Three major factors affect the use and management of molasses grass; ready supply of viable seed, its pioneering potential, and low tolerance to heavy grazing. There are more desirable grasses for long-term pasture use, but the availability of viable seed and ability to grow under difficult conditions make the plant a ready candidate for use in problem situations.

Molasses grass is compatible with most of the perennial tropical legumes, including centro, puero, and greenleaf desmodium. It performs best under rotational grazing or **very** lenient continuous grazing.

Molasses grass can tolerate light to moderate shading, but probably will grow poorly in heavy shade. It grows rather well under young stands of planted pine forests in Hawaii.

Panicum maximum (guinea grass)

An upright tropical bunch grass, native to Africa, guinea is one of the major pasture plants of the tropics and sub-tropics. There are several cultivated varieties. Widely adapted, it is hardy and very productive.

Description. A tall, vigorous perennial bunch grass (Figure 8.18), from 1 to 4 m in height; roots fibrous; culms erect, stout, somewhat flattened; nodes hairy; sheaths smooth or sparsely hairy; ligule fringed with hairs, 1 to 3 mm long; leaves linear, finely pointed, 15 to 100 cm long, 1 to 3.5 cm wide; inflorescence a loose spreading panicle 20 to 50 cm long with numerous erect to spreading branches up to 30 cm long; spikelets without awns, symmetrical, plump, 2.5 to 4 mm long, green or purplish, smooth; seeds 3 mm long, 1 mm wide, somewhat flattened on one side.

Varieties and cultivars of guinea grass. There are a number of cultivars and varieties of this grass. Some are giant upright types; others are small and semi-prostrate. It may be useful to describe some of these.

(1) P. maximum var. trichoglume (green panic). A small, leafy, perennial bunch grass; 0.6 to 1.3 m high, yellow-green in color; leaves 7.5 to 30 cm long, up to 1.4 cm wide; panicles 20 to 35 cm long, lower branches arranged in a whorl; a botanical variety of guinea grass.

(2) Hamil grass. A very tall (up to 4 m or more) variety of guinea grass. Robust, erect, smoother than common guinea, foliage bluer, seeds more freely. Basal leaf sheath with stiff hairs; leaf blades softly hairy.
(3) Colonial. Widely used in Brazil. Giant, 4 m or so high, very smooth foliage, thick stems; foliage blue-green, late flowering.

Propagation and establishment:

Common guinea. Can invade and naturalize lands without sowing or planting. Stool-forming, spreading somewhat by short rhizomes. Can be propagated from seed or crown divisions. Seed viability is often poor. Seeds can be broadcast or drill-sown, but care must be taken so that seeds are not planted deeply -- 6 mm or so is adequate. There are about 1.1 million seeds per kg; planting rate is about 2 to 6 kg per ha.

Vegetative propagation can be accomplished by dividing the basal crown into segments and planting the divisions in furrows or planting holes.

Green panic. Propagated by seed which is commercially available. Seeds are small, about 2.2 million per kg; sowing rate is 0.5 to 5 kg per ha, depending upon viability. Seeds can be broadcast or drill-sown (shallowly).

Hamil grass. Propagated by seed. Follow instructions as for common guinea.

Colonial. Propagated by seed or crown divisions; follow common guinea instructions.

Climatic adaptation. A grass of the medium to high rainfall (875 to 2,500 mm), warm tropics. Because of its deep root system, guinea can withstand quite long droughts, but it grows best under more moist conditions. It is best adapted to coastal areas with good rainfall.

Green panic tolerates low (550 mm or so) to high (1,750 mm or more) rainfall, and is very drought resistant. It grows rapidly in warm weather following rains.

Soil adaptation. Adapted to a wide array of soils, guinea and its relatives nonetheless respond readily to fertilization, particularly nitrogen. Best suited to well-drained soils.

225

Management. As a bunch grass, guinea must be
grazed low enough to keep the plant low and to prevent too much rank, older growth. Grazing height
should usually be about 15 to 22 cm; grazing below
this height may injure the plants and reduce regrowth
potential and even kill the plants. Can withstand
heavy continuous grazing; however, rotational grazing will help to maintain a uniform height for the
pasture.

Legumes which mix well with guinea are centro,
siratro, and leucaena. Legumes should be combined
with guinea in such a way that the legume can grow
in the open spaces between guinea stools to prevent
weed growth as well as to fix nitrogen for the
pasture.

For pastures under coconuts, green panic or
common guinea are probably more suitable than the
tall Hamil or Colonial varieties which are so large
that nut collection would likely be difficult. A
major strength of guinea grass and its relatives is
exceptional shade tolerance. Guinea has been known
to suppress coconut yields under conditions where it
competes with palms for nutrients or moisture.

Figure 8.18. Guinea grass (Panicum maximum).
Culm, inflorescence, spikelet, seed.

226

Paspalum commersonii (scrobic)

A loosely tufted, free-seeding, short-lived, perennial native of Africa, scrobic is widely distributed throughout the moist tropics. It is not as productive as many pasture grasses, but because it can be nearly grazed out by heavy stocking, it may have potential for coconut/pasture systems where intercropping with subsequent food crops on pasture lands may be desired.

Description. A loosely tufted, ascending, short-lived perennial grass; roots shallow; culms erect or ascending, 50 to 100 cm high; nodes smooth; sheaths flattened, smooth except at the hairy mouth near the ligule; ligule short, ending abruptly as though cut off; leaf blades 15 to 50 cm long, 2 to 10 mm wide, flat, smooth to somewhat hairy, margin rough to the touch; inflorescence with 2 to 4 spike-like racemes; racemes spreading, but almost upright, with a few 1 cm long white smooth hairs at the base; spikelets single, alternating in 2 rows along the winged axis of the racemes, nearly round, 2 to 2.8 mm long, smooth, turning blackish brown, stigmas purple.

Propagation and establishment. Propagated by seed which is heavily produced. Sowing rates recommended are 3 to 5 kg per ha; there are about 375,000 seeds per kg. Stands can be maintained by new seedlings.

Climatic adaptation. Best suited to the warm moist tropics, receiving at least 900 mm of rainfall. Mostly a summer growing plant.

Soil adaptation. Can tolerate some waterlogging. Best adapted to fertile soils.

Management. Not very tolerant of heavy grazing. Combines well with siratro and phasey bean.
The Coconut Research Institute of Sri Lanka has studied scrobic pastures. Their results indicate that it is less productive than other pasture plants such as cori and signal grasses.

Paspalum dilatatum (paspalum, dallis grass)

A rapid-spreading perennial from South America, paspalum is adapted to the cooler sub-tropics as well as the warm tropics. It can be used for pasture,

silage, or erosion control, and has become naturalized in many areas.

Description. An erect to ascending perennial
grass (Figure 8.19), spreading by short rhizomes;
culms 40 to 120 cm or more high, usually bent at the
tufted base, sometimes rooting at the lower nodes;
leaf sheaths smooth; ligule a membrane, 2 to 5 mm
long, conspicuous; leaf blade 10 to 20 cm long, 5 to
10 mm wide, midrib conspicuous; inflorescence of 3
to 5, rarely 9, erect or nodding racemes, racemes
widely spaced along the 4 to 10 cm long axis; racemes
4 to 13 cm long; spikelets alternate in pairs in
four dense rows along the rachis of the raceme, the
pedicel of the outer spikelet longer than that of the
inner spikelet of the pair, spikelets fringed with
silky white hairs, light green or purplish in color.

Propagation and establishment. Has become
naturalized in many parts of the tropics and sub-
tropics and comprises a large component of many
"natural" pastures. Propagated mainly by seed.
Sowing rate is 6 to 10 kg per ha. There are about
700,000 seeds per kg. It is best to sow during warm
weather. Can be sown in fresh ash or in cultivated
land.

Climatic adaptation. Adapted to moist sub-
tropics with 800 mm or more rainfall. Is somewhat
drought resistant. Grows best during summer.

Soil adaptation. Widely adapted, but grows best
on soils with good fertility. Responds to N and P;
under declining fertility may become "sod bound."
Can stand waterlogged conditions.

Management. Can stand heavy grazing which helps
to prevent excessive flowering. A naturalizing
grass, it can invade more desirable pastures if
moisture and fertility conditions are favorable. In
most situations, paspalum would not be the grass of
choice for establishment under coconuts.

Pennisetum purpureum (napier grass, elephant grass)

A large, upright perennial bunchgrass from
Africa, napier grass has become an important pasture
and fodder crop. It somewhat resembles sugar cane.
The most productive tropical forage grass, it can
yield as high as 100 metric tons of green forage per

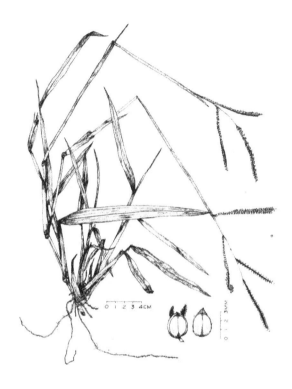

Figure 8.19.   Paspalum (Paspalum dilatatum).
Plant habit; spikelets, 2 views.

hectare per year; to produce high yields, high levels of N are required.

Description. A tall, robust perennial bunch-grass (Figure 8.20); deeply rooted, spreading some-what by short creeping rhizomes to form large clumps or stools; culms 2 to 7 m high, stout, generally smooth, nodes hairy or hairless; sheaths smooth and hairless or rough and hairy in upper parts; ligule a very narrow rim, fringed with dense white hairs up to 3 mm long; leaf blades 30 to 90 cm long, up to 2.5 cm wide, stout midrib; inflorescence a compact, erect, cylindrical spike, 8 to 30 cm long, 1.5 to 3 cm wide, yellow or tinged with brown or purple; spikelets arranged around a hairy axis, falling with bristles attached at the base from the axis at maturity, each spikelet surrounded by numerous bristles; spikelets with 2 florets, lower floret male or barren, upper floret usually perfect.

Propagation and establishment. Propagated by stem cuttings, much like sugarcane. Cuttings with 2 to 3 nodes survive best. Cuttings can be planted in furrows or planting holes in rows, and covered with soil. Can become naturalized in uncultivated lands.

Climatic adaptation. Grows best in warm, moist conditions, where rainfall exceeds 1,000 mm. Al-though it is best adapted to more humid conditions, it can survive moderate drought because of its deep root system.

Soil adaptation. Adapted to a wide range of soils, from sands to well-drained heavy clays. Grows best in well-drained fertile soils and under high rainfall. Cannot withstand waterlogging. Very responsive to irrigation and nitrogen fertilizer. Has a high K requirement.

Management. Best used as a cut fodder or soilage crop; difficult to graze properly because the stems become rank, fibrous, and of low quality and palatability. May be necessary to cut or slash down periodically to 15 to 30 cm height, to keep growth fresh and palatable. Can be used for silage or green feed. Can be used as a special-purpose fodder crop to produce high yields of forage during periods of feed shortage. An excellent grass to use under irrigation and with N fertilizers in intensive production systems for small farms.

Combines well with several legumes including calopo, centro, puero, and glycine, when grass and legume are planted in alternate rows.

For coconut lands, unless carefully managed, napier may become too tall for nut collection. It would probably be best to manage it as a cut green forage, rather than as a pasture under the palms. Also, because napier has a high N and K requirement, it could compete heavily with coconuts for these nutrients.

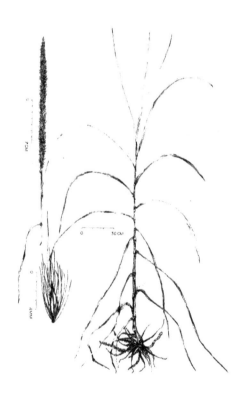

Figure 8.20. Napier grass (<u>Pennisetum purpureum</u>). Plant habit, inflorescence, spikelet.

9
Managing Cattle Under Coconuts

Flemming I. Ericksen and D. L. Plucknett

GRAZING MANAGEMENT

Grazing management objectives are basically the same under coconuts as on conventional pastures; these include (1) maintain high production of good quality forage for the longest possible time, (2) maintain a favorable balance between forage species, (3) encourage rapid regrowth during and/or after grazing or cutting, (4) make wise compromises between yield and forage quality, (5) achieve efficient utilization of the forage produced, (6) obtain high animal production, and (7) minimize the need for mowing, weed control, or other costly operations. In the following sections some of the above-mentioned points will be discussed in greater detail.

Before going on to a discussion of those objectives, however, it may be useful to discuss briefly some of the special considerations of grazing cattle under coconuts.

Cattle Damage to Coconuts

It is commonly accepted that cattle can cause damage to young coconut palms. Numerous authors (Salgado, 1953; de Silva, 1953; Ferguson, 1907) have called attention to damage caused by stray cattle; in Sri Lanka stray cattle have been termed the worst pests of coconut (de Silva, 1953). In areas where little forage or feed is available, such underfed uncontrolled cattle will eat the new, tender, inner leaves of young palms, (Figure 9.1), setting them back and so weakening them that they die, or at best are unproductive. Constant

Figure 9.1   Cattle damage to a young palm in
            Western Samoa. Note browsing of
            leaflets and absence of new growth
            in the center of the palm; new leaves
            have been grazed off.

233

defoliation leads to tapering of the trunk, and
the poor start often reflects throughout the life
of the palm (de Silva, 1953). If the terminal
growing point is damaged, the plants will die.
When the trees become larger -- at about four to
six years of age -- the young inner leaves usually
are beyond the reach of cattle, and browsing
damage is no longer a problem. It has been recommended that cattle be kept away from young palms
for at least three years after planting (Ferguson,
1907). de Silva (1953) suggested several methods
of protection for the young palms. These included;
fencing of individual plants or concentrated plantings, spraying a slurry of manure on the leaves,
and use of chemical repellants.

Some authorities have suggested using weaner
cattle or sheep to graze in young palm groves, because presumably they do not browse on young palms
(Munro and Brown, 1916; Frank Moors, pers. comm.;
Morris Lee, pers. comm.). Weaner cattle are
grazed successfully in 4 year-old palms in Western
Samoa; the main factor to consider is whether the
pasture and palms have been fertilized to ensure
a good feed supply for the animals as well as good
growth of palms. Frank Moors (pers. comm.) states
that unfertilized young palms should not be grazed
until 6 years of age. This has been confirmed in
the Solomon Islands where fertilized, young palms
established by an improved polyethylene bag planting technique can be grazed successfully within
three years (Ian Freeman, pers. comm.). However,
not all persons agree on this point; for example,
R.A. Williams of the Coconut Industry Board in
Jamaica (pers. comm.) recommends waiting until
palms are 7 to 9 years old before turning cattle
into the groves.

Indirect damage to coconuts by cattle results
from trampling, soil compaction, and soil exposure
near the trees (de Silva, 1953). Such damage can
result in poor root penetration and development, as
well as   soil erosion and loss of nutrients.
Provision of a good cover crop can help to reduce
such losses. When overgrazing occurs, cover crops
become weakened and the palms again suffer.

Production of good pasture forages may be the best way to prevent cattle damage to coconut. In pine forests in the southeastern U.S.A., cattle damage to young trees is held to a minimum by establishing a good forage stand and by moderate grazing pressure to avoid overgrazing and degeneration of the pastures (Plucknett and Nicholls, 1972).

Maintain High Production of Good Quality Forage for the Longest Possible Time

In order to maintain high production of good quality forage for the animals, the pasture must be grazed in such a way that overgrazing, or undergrazing, do not occur.

If an established pasture is grazed very heavily, very few leaves will be left for photosynthetic activity, and the plant will have to draw heavily on its food reserves, carbohydrates that are stored in the plant roots and crowns to produce new foliage.

The general rule is that if plants are grazed too heavily, they must have a long rest period before being grazed again, to build up plant carbohydrate reserves. If the plants are not given enough time to recover before the next grazing, they will continue to draw on their carbohydrate reserves; the roots will decrease in size, weight, and number; the plant will become more susceptible to droughts; recovery after grazing will be slower; and the plant may eventually die, with the result that less desirable plants will take over.

With light to moderate grazing, however, sufficient green leaf area remains after grazing, the plants do not draw heavily on their carbohydrate reserves, and therefore the pasture may be grazed more frequently. Light or moderate grazing is especially essential for the persistence of tall bunch grasses and most legumes, whereas most prostrate, stoloniferous grasses are able to withstand closer grazing.

Undergrazing will lead to selective grazing. The animals will only eat the most palatable plants and leave these that are less palatable or less nutritious. Over the long term, selective grazing leads to a decrease in the total quality of the

pasture. If some pasture plants have grown too old
and stemmy, the pasture must be mowed or slashed,
to allow for new growth with increased nutritive
value and greater palatability.

Maintain a Favorable Balance Between Forage Species

Proper management of a pasture will lead to a
favorable balance between forage species. If a
grass-legume pasture has been established, it is
recommended to graze the pasture to suite the legume, because legumes are usually the most difficult to manage. Overgrazing of a grass-legume
pasture usually leads to a fast decline in the
legume stand and an increase in weed population.
Tropical kudzu (Pueraria phaseoloides) is an
example of a good legume that can be easily grazed
out.

If the legume stand in a grass-legume pasture
is poor, it is usually possible to increase the
stand by leaving the area ungrazed for a while to
reseed or spread, and/or by fertilizing with phosphate and perhaps potash. In a grass pasture where
weeds are dominant it is usually possible to increase the grass stand by leaving the area ungrazed
for reseeding (if viable seed is produced), or by
fertilizing with nitrogen fertilizer. Most improved grasses respond readily to nitrogen fertilizer and can then compete better with weeds.

Encourage Rapid Regrowth During And/Or After Grazing or Cutting

Rapid regrowth can only be expected if the
pasture has not been grazed too close. Some green
material must be left after grazing to ensure rapid
regrowth.

Under continuous grazing where cattle stay in
the same paddock all the time, they will selectively graze the young, most palatable plants. A few
days later the cattle will graze the new shoots of
the same plants again. After each time the cattle
graze that same plant, regrowth will be slower.

Rotational grazing has some advantages because
the pasture is grazed heavily and more evenly within a short period of time. After the animals are
taken off, the pasture is left ungrazed until the
plants have restored their carbohydrate reserves.

Through a well-planned rotational grazing schedule we can expect regrowth after grazing.

Application of fertilizer after grazing will encourage rapid regrowth.

Make Wise Compromises Between Yield and Quality

The maximum yields of dry matter are obtained when the pasture is harvested at or near maturity, but the nutritive value and digestibility of the forage are low at this stage, due to high crude fiber and low protein content. On the other hand, the nutritive value is highest in young leafy growth due to high crude protein and minimum crude fiber content. Figure 9.2 shows the general relationships between digestibility, crude protein, crude fiber, and total yield for a representative tropical grass at different cutting intervals.

The right time to graze a pasture varies with factors like species used, time of the year, soil fertility, moisture condition, and types of animals (e.g. milking cows and young growing animals require more protein than dry stock).

Management should be regulated to prevent flowering of most pasture plants, particularly grasses, because flowering reduces nutritive value and palatability of the forage. If flowering cannot be prevented by grazing, the flowering stems should preferably be slashed or mowed.

High Animal Production by Efficient Utilization of the Forage Produced

The farmer aims for the highest economic output per ha from his pasture. To learn how to obtain the highest output with time may require many years of experience. One reason is that pasture production varies from year to year, mainly due to variation in rainfall and growing conditions. A rule of thumb is not to graze more animals than the poorest year can carry, unless supplemental feeding is available. The reason for this is that overgrazing for a short time might damage a good pasture to the extent that costly reestablishment or rehabiltation may be required.

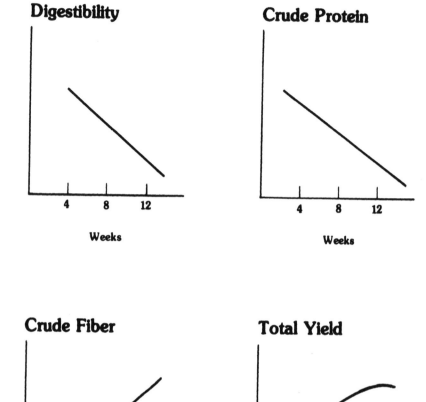

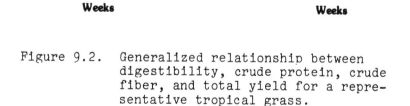

Figure 9.2. Generalized relationship between
digestibility, crude protein, crude
fiber, and total yield for a representative tropical grass.

Grazing pressure on a pasture significantly affects animal performance. The highest output per individual animal is obtained under low grazing pressure, because animals graze selectively and only feed on the most palatable, most nutritious parts of the pasture. Because of selective grazing, a part of the pasture will not be utilized and will become rank and useless, resulting in needless waste of forage produced.

In certain situations undergrazing may be desirable. For example, when steers are being finished on pastures, undergrazing (light or moderate grazing) might be useful because a high daily gain is important for a good classification of the carcass. After the feeder steers are moved from a paddock, unutilized forage can then be consumed by a herd requiring lower quality feed or where high daily gain is of less importance, such as in dry stock.

As the number of animals is increased, selective grazing is diminished since the fixed amount of forage available must be shared by more animals, and the daily individual animal gain is decreased. However, in such a situation, total production per hectare will usually be higher than under light or moderate grazing. The optimum number of animals per unit area therefore falls within a range, which is a compromise between gain per individual animal and total production per ha.

A further increase in stocking rate into the "overgrazing" range results in only enough forage consumption per animal to meet individual maintenance requirements, and gain per animal becomes zero or even negative. In such a situation, both individual animal performance and total production per ha decline, and pastures can be damaged seriously.

Recommendations for Grazing During the Dry Season or When Pasture is in Short Supply

When a dry season comes, stock suffer feed shortage, lose condition and body weight, and -- in very severe cases -- may even die. For some areas a dry season is normal each year, and farmers should be prepared to meet the problems that will be faced. Many of the potential practices that can be employed have been covered to some extent in this chapter and in Chapters 5 and 7, but one topic that has not been covered adequately is the priority of grazing of the most nutritious forages among various classes of cattle. Osborne (1978) recommended that priority in nutritional well-being be assigned (from highest to lowest) as follows: (1) weaner heifers, (2) cows in advanced pregnancy, (3) heifers during first pregnancy, (4) other pregnant cows, (5) non-pregnant breeders, and (6) two-year-old unmated heifers and other dry stock, including steers and culls.

SYSTEMS OF GRAZING

Continuous Grazing

Continuous grazing is an extensive system of grazing in which the livestock remains on the same pasture year-long. While continuous grazing using low stocking rates may prove as productive as rotational grazing, tropical grasslands are normally undergrazed during the rainy season and overgrazed during the dry season, with consequent deterioration of the sward and possible soil erosion at the beginning of the rainy season.

Where continuous grazing is moderate, it is perhaps more economical than rotational grazing, especially when; (1) construction of fences or barriers is very costly, (2) access to water holes or waterlines is very difficult or costly, and (3) the system is used on small farms with only a few animals.

A disadvantage of continuous grazing, as mentioned earlier, is the possibility of selective grazing which usually leads to a deterioration of pasture quality and production capacity (Goonasekera, 1951: Goonasekera, 1953a, 1953b; Rajaratnam and Santhirasegaram, 1963; Schrader, 1967). With selective grazing we can expect some growth of shrubs and some poorly grazed areas, which will make it more difficult for the coconut-collectors to find the nuts. Also, it is easier to collect the nuts if no animals are in the paddocks during collection operations.

In Sri Lanka it was found that coconut palms suffered severely from competition for nutrients and soil moisture with large shrubby weeds, when continuous grazing was practiced (Goonasekera, 1951).

Another serious disadvantage of continuous grazing in the tropics is the build-up of tick and nematode infestation. If young animals are grazed continuously on the same pasture as older animals, they become heavily infested with helminth parasites and their growth is retarded. Under a system of rotational grazing the degree of infestation is considerably reduced (McIlroy, 1972).

Rotational Grazing

Rotational grazing is an intensive system of pasture management. The grazing area is subdivided into a number of paddocks, and the animals are moved systematically from one to another of these in rotation. Paddock size depends on: (1) number of animals in each herd division, (2) access to water, (3) cost of fencing material, (4) estimated number of days between grazings, and (5) age and spacing of coconut trees.

The grazing interval for pastures in the humid and sub-humid tropics ranges from 30 to 60 days, depending on factors like pasture species used, class of animals used, time of the year (rainy or dry season, temperature), and, if fertilizer is used, the amount and when it was applied. Fortunately, the interval between coconut harvests falls within the 30 to 60 day range, or could fit in with multiples thereof. Ideally, nuts should be collected at least monthly. Gathering fallen nuts immediately after grazing would make it very easy to find and harvest the nuts (Javier, 1974).

The length of the grazing period depends on the stocking rate and forage growth rate per unit area. When the animals have grazed one paddock down (a height of 7 cm is often recommended as the rule of thumb), the animals are moved to the next paddock while the first is rested. Often a three or four paddock system is used in this way. By the time the last paddock in the sequence has been grazed down, the first paddock should be ready for grazing for the second time. The system is based on the assupmtion that; (1) the greatest herbage production is obtained when the pasture is rested before the next grazing, (2) that animals in large numbers on a small land area are forced to spread over the entire area to use the available forage more uniformly, and (3) that trampling is reduced because animals are held on small areas where feed is more abundant and hence, where less travel is necessary.

Most commonly, rotational grazing is practiced by using one group of cattle. It is possible though, to obtain a more efficient utilization of the available pasture by dividing the grazing animals into two groups, high producers (high-producing milk cows, steers, or calves where fast gains are desired) and low producers (dry cows and stores). The high producers first graze the pasture for a short period, and are then followed by the low producers, who consume the remainder of the less nutritious forage that remains.

Very little work has been published on the comparison of continuous versus rotational grazing under coconuts. In an experiment in Sri Lanka Rajaratnam and Santhirasegarem (1963) compared the herbage yield of Brachiaria brizantha under rotational and continuous grazing, and found that the yield was slightly higher under rotational grazing. In several other reports, rotational grazing under coconuts is recommended (Eden, 1953; de Silva, 1961; Goonasekera, 1951; Javier, 1974).

In Sri Lanka, Ellewela (1956) recommended rotational grazing with no more than one animal per 0.8 ha on unimproved pastures. Remalingam (1961) found that yields from Brachiaria brizantha pastures did not differ under rotational and continuous grazing.

Disadvantages with rotational grazing are the increased cost of fencing, higher labor cost involved in moving the cattle, and increased management skills required.

Deferred Grazing

Deferred grazing is the practice of delaying grazing during part of the growing season as a means of conserving forage until it is needed, and also as a means of pasture improvement. In most

242

parts of the tropics plant growth is uneven during the year, due to uneven rainfall distribution. Under a system of rotational grazing, certain pasture paddocks with abundant plant growth can be set aside during part of the year to be used later, usually during the dry season. Of course, it must be remembered that the feeding value and palatability of an over-matured pasture is very low. Such a pasture generally supplies little more than a maintenance ration, but it can be of great value to carry animals over the peak of the dry season. Guinea grass (Panicum maximum) and stylo (Stylosanthes guyanensis) are examples of forages that could be used in deferred grazing.

The purpose of deferred grazing as a means of pasture improvement is to build up plant vigor, to allow the root system to develop, and to allow self-sown seedlings to become established.

A disadvantage of deferred grazing under coconuts will be the difficulty of collecting nuts in an ungrazed pasture. Pastures destined for a deferred grazing system should probably be located in unused, open areas near the groves or in scattered, old, or senile coconut stands that have not been underplanted.

Zero-Grazing

Zero-grazing refers to feeding housed, penned, or tethered stock with freshly cut crops. Instead of bringing the animals to the pasture the pasture is brought to the animals. Advantages of zero-grazing are: (1) no losses from fouling or trampling, (2) higher animal production owing to more efficient forage utilization, (3) higher forage production due to harvest at a more advanced stage of growth, (4) animals can be protected from heat in the open, and (5) no expenses for fences and piped water.

Disadvantages of zero-grazing include: (1) cost of equipment needed for cutting and carting the forage to the animals, (2) very labor intensive, (3) extra cost of housing, pens, or yards, (4) manure disposal may present a problem, (5) difficulties in cutting forage crops between the coconut trees, and (6) damage to fallen coconuts during harvest.

Zero grazing is a very intensive system. The species recommended would be high-yielding, tall plants like guinea grass and napier grass (Pennisetum purpureum). Because these grasses compete with coconuts for nutrients, fertilization of coconuts would be required.

Zero-grazing could be practiced during the dry season when pasture production is low. Pastures left ungrazed during the wet season can be cut and the feed brought to the animals. This method would increase forage utilization, because waste due to trampling can be extremely high when a pasture is grazed at an advanced stage of growth or during the wet season.

Tethering

Tethering of cattle to palms in order that they may graze on natural or improved forages around the trees is a common practice in many coconut areas (Figure 9.3) (Goonasekera, 1951; Piggot, 1964; Salgado, 1947; de Silva, 1951). Tethering of cattle to palms is practiced particularly on small-holdings where only a few cattle are kept, and therefore it would not be economical to fence off the area. Tethering can be practiced on large farms too, if labor is cheap, or if village or family cattle are grazed by mutual agreement under the palms. Goonasekera (1951) described how a four hundred hectare farm in Sri Lanka was rotationally grazed by tethered Sinhala cattle. Each of the animals was tethered separately to individual palm trees with a rope allowance equal to the distance (8 m) between the palms (Figure 9.4). Each square of palms was thus grazed down evenly to 10 cm, after which the animal was moved to another palm tree.

Tethering of cattle around coconut trees at night as a means of manuring has been used extensively in Sri Lanka (Schrader, 1951). Two heads are tethered for ten nights to a palm, round which a shallow trench has been dug; after the tethering period the trench is filled in. The number of palm trees manured in this fashion is rather low, about 18 trees per head of cattle per year.

Night-penning

Night-penning is used to provide for improved

Figure 9.3.   Tethering of animals to palms can be a very useful and low-cost method of controlling animals, while at the same time utilizing the forage. Above: (a) water buffalo, Philippines; (b) dairy cow (Sinhala), Sri Lanka; (c) cow, Philippines; (d) goat, Sri Lanka).

Figure 9.4. Cattle tethered to palm in Sri Lanka,
on unimproved pasture. The rope
length is enough to reach the distance
to the nearest palm. Below: the animal has been tied shorter than usual
to prevent grazing of the young underplanted palms.

production of farm-yard manure, and to avoid un-
necessary trampling of the pastures.

A system of manufacturing farm-yard manure by
night-penning in Sri Lanka was described by
Schrader (1951). If the supply of straw was abundant, the quantity of manure produced could average
as much as 10 metric tons/head/year, which will
supply manure for 200 palms. One should be aware
of the fact, however, that manure pits are excellent breeding places for rhinoceros (<u>Oryctes</u>)
beetles. On the other hand, when adequately
managed, the eggs and larvae will be destroyed and
the pit may thus function as an insect trap
(Fremond, 1966; Schrader, 1967; de Silva, 1951).

Based on practical experience, Munro and
Brown (1916) stated that 150 head of cattle and
sheep (the number of each type of animal was unspecified) could produce enough manure each year
for 20 ha of palms.

Trampling can cause soil compaction on heavy
clay soils. Since cattle graze mainly during day-
time (Goonasekera, 1954) the cattle can be taken
into a corral at night, without affecting their
food intake, especially when some extra rations are
fed in the corral. Extra grazing at night or extra
feed should be provided if the daily grazing has
been disturbed because of heavy rain.

Stocking Rate or Carrying Capacity

The stocking rate or carrying capacity of
pastures under coconuts may vary according to
differences in: (1) planting density of the palms;
(2) botanical composition of the pasture; (3)
climatic factors (temperature, total rainfall, and
annual rainfall distribution);(4) soil fertility
and amount of fertilizer used; (5) type of cattle
(or other livestock) used; (6) grazing system
used; (7) additional rations fed to the cattle;
and (8) the efficiency of management of the available resources on the farm.

The term, carrying capacity, is not very well
defined. It is more appropriate to use the term,
animal unit, rather than number of animals carried
per ha, which is the way carrying capacity is often
expressed. The "animal unit" equivalent system is

used as a guide to determine the carrying capacity when several classes of animals are grazed together, or when one class of cattle is substituted for another (Table 9.1).

Carrying capacities reported from different parts of the tropical world vary widely. In Sri Lanka, carrying capacities of 1.25 to 5 heads of the small native Sinhala cattle per ha are reported (Burgus, 1938; Ellewela, 1956, 1957). In some cases in Sri Lanka, cattle are fed "poonac" (coconut press cake) at about 1 kg per head as a daily supplementary feed, and this would certainly have an impact on the number of animals carried on a pasture.

Table 9.1. Animal unit equivalents for the various classes of beef cattle (Vallentine, 1965).

Class of Cattle	Animal Units
Mature cow, maintenance	1.00
Mature cow with calf, birth to 3 months	1.25
Mature cow with calf, 3 months to weaning	1.40
Weaner calves, to 12 months	0.50
Yearlings, 12 to 17 months	0.65
Yearlings, 17 to 24 months	0.80
Two-year-old steers	0.90
Bulls	1.25

Rajaratnam and Santhirasegaram (1963) found that an adult Sinhala cow could be maintained on 0.15 ha of guinea grass, while 0.5 and 0.55 ha, respectively, were necessary in the case of cori grass (Brachiaria miliiformis) and palisade grass (Brachiaria brizantha). In the Ivory Coast, about 0.75 head per ha could be maintained under coconuts on sandy soils covered with centro (Centrosema pubescens) (Ferdinandez, 1968). Under favorable conditions in the New Hebrides, Solomon Islands, and Western Samoa, it is possible to carry up to 3 head per ha (Eden, 1953; Fremond, 1966; Groeneweld, 1967). In Western Samoa the liveweight gain recorded for steers stocked at 2 to 3 head per ha averaged 450 kg/ha/year on guinea grass/centro pasture, and 150-200 kg/ha/year on good unimproved pasture (Ericksen, 1973; Reynolds and Schleiker, 1975).

The production of yearling Bali cattle (Bos banteng) grazed on improved pasture under coconuts ranged from 302 kg liveweight gain/ha at a stocking rate of 2.7 yearlings/ha, to 611 kg/ha at a stocking rate of 6.3/ha (Nitis et al., 1976).

No general recommendation can be given for optimum stocking rate on a particular pasture; only years of experience will tell the farmer how many animals he can keep on the farm. Paltridge (1956) says, "The optimum and the correct stocking rate for any pasture is that number of animals, which can obtain adequate feed (on a year-long basis) without serious detriment to the pasture". Only through good year-round management is it possible to obtain maximum output. For example, under conditions with a dry and a wet season it is possible to maximize the stocking rate by providing extra rations of concentrates to the cattle during the dry seasons. If a rotational grazing system is used, some pastures can be left ungrazed during the wet season to be used during the dry season (deferred grazing), or it might be possible to make hay or silage during the wet season for use during the dry season. Also, it may be possible to plant specialized pastures in swampy areas or otherwise unutilized land, to ensure a year-round feed supply.

If a pasture has been undergrazed, a lot of old stemmy plant material will be left over. It is important that the stemmy material be removed to allow new regrowth. One way of doing this is to allow a very high stocking rate for a short time, which will force the animals to eat the stemmy material; this method is generally called "brushing". Due to the poor quality of the pasture, only store cattle would be recommended for brushing. Brushing is often used also before land preparation prior to new seeding or planting.

SUITABLE ANIMALS

Breeding and Benefits of Shade

European breeds (Bos taurus) have been introduced to the tropics, but due to the high temperatures, solar radiation, and humidity the production of these breeds is lower in the tropics than in the temperate zone. Several strains of cattle indigenous to the tropics have been selected and used successfully. An especially successful group has been some select breeds from India belonging to the species, Bos indicus. Bos indicus cattle are characterized by a prominent hump located over the withers. There are Asian and African types. In the following discussion the term "zebu" will be used synonomously with B. indicus.

Heat has a greater effect on the well-being of cattle than any other climatic factor. High ambient temperatures cause an increase in body temperatures and increased respiratory rates. Rising body temperatures suppress metabolism to the extent of inducing a disinclination to move, loss of appetite, and reduced production. Johnson et al. (1953) recorded that for Holstein cattle a 1° C increase in rectal temperature caused an approximately 3.3 kg loss in milk production, and a 2.5 kg loss in consumption of total digestible nutrients. Kibler and Brady (1950) state that rectal temperatures of European cattle rise at atmospheric temperatures between 21 and 27° C; in the zebu this is not observed until ambient temperatures of 32 and 34° C are reached.

The reasons for the superior heat-tolerance of the zebu are still not clear. Their lower milk yield, slower maturity as regards weight and reproduction, as well as longer intervals between calves suggest a lower rate of metabolism (Sanders, 1966). Many breeds indigenous to the hot humid tropics are small or even dwarfed. Such animals would have a good surface area/mass ratio as an aid to heat dissipation; the long ears and loose pendulous dewlap of the zebu would serve this purpose. Another advantage of the zebu breeds is their short, thick medullated hair, compared to the long fine hair of European breeds in the tropics.

Zebu cattle are known to have a greater resistance to a number of diseases and parasites, and also have a longer productive life than European breeds in the tropics.

Crossing European and zebu cattle to produce new types of beef and dairy cattle has been very successful. It must be remembered that the tropical environment varies so much that no single breed will meet all conditions, and the final choice of animal will depend upon the local circumstances. The Santa Gertrudis breed, which is three-eights zebu, five-eights Shorthorn, is just one example of a successful cross breed used for beef. Santhirasegaram (1967) recommended a half-breed Sinhala-Jersey cow as the most suitable dairy animal for the low country

250

wet zone (Coconut Triangle) of Sri Lanka.

Whatever breed is used, serious efforts must be made to reduce at least some of the stresses of the tropical environment by disease and pest control, pasture improvement, and provision of shade or housing for protection from high atmospheric temperatures, direct solar radiation, and high rainfall.

Coconut groves provide an ideal partial shade for cattle. Observations from Sri Lanka (Goonasekera, 1953) have shown that the indigenous low country Sinhala cattle graze for the longest period during the hottest part of the day, from noon to 6:00 p.m. Eden (1954) reported from Western Samoa that Hereford and Holstein breeds do not graze during the hottest time of the day. Personal observations from Western Samoa (F. I. Ericksen) show that Hereford-zebu cross do graze during the hottest part of the day under coconut, but not on open land.

High environmental temperatures in the tropics, coupled in some cases with high humidity, are apparently the main cause of low productivity, especially among European breeds. This high temperature effect is reduced by grazing cattle under coconuts. Air temperature measurements in the Philippines under the coconut canopy during sunny and cloudy conditions show a difference of at least 6 degrees C during the day from 9 a.m. to 1 p.m. (de Guzman, 1975).

HERD MANAGEMENT

Not many publications on livestock (especially cattle) management have been written with the small producer in mind. For good examples of recent publications for small farm cattle and forage production, see: (1) Beef Production Manual, University of the Philippines at Los Banos, (1974); (2) Philippine Council for Agriculture and Resources Research, (1976); Plucknett, (1978); Small Beef Herd Manual, Michigan and Pennsylvania Agricultural Experiment Stations).

Controlled Mating and Calving

Year-round mating is a common practice in many places. The reason for this could be that conditions are so variable and unpredictable that immense losses could be sustained through low conception rates at any one particular mating time. Also, perhaps in a small herd the lack of fencing makes it impossible to segregate the herd and to separate the bulls. In areas where pasture production is evenly distributed during the year and where an even supply of meat for the market is needed, year-round mating might give the best output. For a dairy herd, year-round milk production is preferred, which can only be obtained by year-round mating.

In most areas, pasture growth varies during the year and it would be recommended to arrange for calving at the start of the expected flush in feed supply (early rainy season), to assure calves the maximum number of months of good feed. Controlled mating in a beef herd also concentrates routine tasks such as branding, ear-marking, dehorning, castration, vaccination, and weaning into short periods instead of having them drag on year-round.

Another advantage of calving at the beginning of the flush in feed supply is that the rapidly-growing young pasture has a high protein content which is ideal to build up a high level of milk production in the cows. Secondly, this feed is also ideal for young calves which have yet to develop the rumen capacity for the more fibrous feed which will come later. Thirdly, the best feed period of the year will coincide with the time, about 2 months following calving, when the cows -- already at the height of their lactation -- should be re-mated.

The duration of the mating period is a compromise between the benefits of compressing calving time into a defined time of the year, and the need for a long enough mating period to ensure a high conception rate. Ten weeks, which allows enough time for 3 complete oestrus cycles, is recommended under good conditions of management and nutrition.

Under a system of controlled mating the proportion of bulls used may vary from 3 to 6% (3 to 6 bulls per 100 females) of the herd. Only three percent may be needed where stocking rate is high and the terrain is even, while more bulls are needed in paddocks of low stocking rate and rough terrain, and if the bulls are very old or very young.

At what age should the heifers be mated? To obtain high productivity, it is important that the heifers are raised under good conditions, so they can reach the mating stage early. Liveweight is more important than age as a criterion of eligibility of mating. If only one calving period per year is adopted, it is important that the heifers reach the required weight for breeding by 15 months, or they will have to wait until the next mating season when they will be 27 months of age.

Pregnancy testing about 8 weeks after the completion of mating is desirable as an aid to culling and pasture management.

Weaning

The age at weaning -- separating cows from their calves -- can vary, depending on the amount of milk being provided by the cows. In general, 6 to 8 months is the average weaning age, but if the cow is rapidly losing condition or when drought is experienced, it would be advisable to wean at an earlier age.

Weaning represents both a nutritional and a physiological change. Before weaning, the calf has been constantly associated with its mother and has depended on her for both protection and food. Suddenly it is cut off from both and is naturally nervous and under stress, until it learns to readjust to its new environment. Further stresses at weaning may also be caused by vaccination, drenching, dehorning, weighing, and castration. To minimize weight loss immediately following weaning, the weaners should be placed on the best pastures. Bull calves which have been selected as replacement stock should be segregated as early as possible, to safeguard against accidential pregnancy of immature heifers. Steer and heifer calves can be allowed to graze together, providing there is no wide variation in age. If possible, a good practice would be to graze the weaner calves ahead of the adults on paddocks used on a rotational basis. This should not be difficult with coconut pastures.

If possible, it would be a very valuable routine to weigh the calves at weaning, as a part of a performance testing program that will provide a more objective basis for the culling of unwanted animals.

Culling

A good time to consider culling is at weaning, when all the cattle are together. The first group to be considered for culling are the cows which were not pregnant at the time of pregnancy testing. If the annual intake of good heifers into the herd is more than sufficient to balance the number of non-pregnant cows, the non-pregnant cows should be culled. If there are more unbred cows than can be replaced by heifers, some of the unbred cows with the best past record or future potential may be retained. In a situation with a very high percentage of non-pregnant cows, a second mating season might be considered, instead of keeping the unbred cows until the next annual mating. Another advantage of having a second mating season would be that heifers not quite heavy enough for mating during the first mating season could be mated during the second mating period. A second mating could be envisaged only in regions with fairly even rainfall distribution, where feed supply is fairly uniform throughout the year.

Another group to be considered for culling are the poorest of the heifer weaners, measured on a weight-for-age basis, and mothers of such poor calves. Again, a balance has to be kept between the intensity of culling and maintenance of herd numbers at the desired level.

A farmer might want to cull his cows at a predetermined age. There is justification for culling by age when no means exists to identify good and poor producers in terms of their calves. However, when the calf/dam identity and performance are known, a far better procedure is to cull each cow when it reaches the stage of producing a calf which will be inferior at weaning to the calf likely to be produced by the heifer that will replace the cow. The same can be said of culling cows on a visual basis for various "faults", like weak legs or udder defects.

The criteria for culling mentioned here are for beef cattle. For dairy cattle some of the criteria would be the same, like culling for infertility. If a dairy cow does not drop a calf every year, milk production will not be acceptable and the cow should be culled. The most important criterion for culling dairy cows would be on the basis of individual annual milk production. Other important criteria that could be mentioned include udder form and ease of milking.

Fertility

The most important factor for successful cattle production is a high fertility rate in the herd. Factors contributing to infertility are: (1) nutritional, (2) genetic, and (3) infectious diseases.

To obtain a high conception rate during mating, the animals bred must be on a plane of increasing liveweight gain, and mating should coincide with a period of minimum stress. If the pasture production and quality are low, supplementary feeding during

254

mating might be necessary. Fertility may also be affected by the low phosphate status of most tropical soils, a problem which is usually reflected in the pasture forage produced.

To be sure the cattle get enough minerals, a mineralized salt containing phosphate and other important elements must always be available for the cattle to eat.

It is most desirable for animals to conceive soon after calving, and this may be related to genetic factors. Bos indicus breeds may manifest poor fertility, and it is important to practice selection to combat this defect.

Venereal diseases -- as well as contagious abortion -- also adversely affect fertility, but vibriosis is the most common cause of infertility in most countries.

Artificial Insemination

Artificial insemination (AI) is, by definition, the deposition of spermatozoa in the female genitalia by artificial rather than by natural means.

Some of the advantages of AI are:

(1) It increases the use of outstanding sires. Through AI, many breeders can use an outstanding sire, whereas the services of such an animal were formerly limited to one owner, or -- at the most -- to a partnership.

(2) It makes it possible to use breeds which were not available otherwise. Many countries do not allow importation of live cattle from other countries, in order to prevent diseases from coming into the country. Through the use of imported frozen semen for use on the best of the local cattle, the herd can be improved without the danger of introducing any parasites and diseases.

(3) It may lessen and control certain diseases. AI may eradicate or prevent the spread of certain types of cattle diseases, especially those associated with the organs of reproduction, such as trichomoniasis, vibriosis, and vaginitis. However, when improperly practiced, it may be an added means of spreading disease. Therefore, it is very important that all males used be examined carefully for symptoms of transmissible diseases, that bacterial contamination be avoided during the collection and storage of semen, and that clean, sterile equipment be used in the insemination.

(4) It lessens sire costs. On small farms with only a few cows AI would usually be less expensive

255

than the ownership of a worthwhile bull.

Artificial insemination is not without limitations; these include:

(1) It requires skilled technicians for insemination, and skilled management to detect the heat period. In order to be successful, AI must be carried out by skilled technicians who have had considerable training and experience. AI can only be successful if done at the right time of the oestrus cycle. Good heat period detection depends on adequate observation at the proper time. Cows are most active in the early morning and near dark, which is the best time to check the herd to find the cows which are in heat.

(2) It requires considerable capital to initate and operate. Considerable money is necessary to initiate an artifical insemination program, and even more is needed to expand and develop it properly. In some countries, government agencies provide AI services for small farmers.

DISEASES AND DISFUNCTIONS OF THE GRAZING ANIMAL

Diseases found in the tropics are most of those found in temperate climates, plus many others. Complete accounts of the diseases of cattle are readily available (e.g. Hungerford, 1962), so no attempt will be made to provide a full coverage. Horng (1976) presented a list of diseases in the Asian tropics where the largest share of the world coconut crop is located.

SUPPLEMENTAL FEEDING

Most areas experience seasonal pasture growth, such that there is an abundance of feed for some months of the year and insufficient feed during the remainder. Supplementary feeding during the part of the year when pasture is lacking or of poor quality can be used to overcome feed shortages. In a dairy herd supplementary feeding would also be recommended during the peak of milk production, to assure that the cattle get sufficient energy and protein for optimum milk production.

When to use supplementary feeding, how much, and to what animals depends mainly on the availability and price of the feed supplement. After all, a farmer will only use supplementary feeding if he gets a higher profit by doing so.

In a beef herd with annual mating the priority of supplementary feeding would be to: (1) weaner stock if high quailty pastures are not available, (2) those heifers approaching 15 months of age that appear likely not to achieve sufficient weight for mating on pasture alone, (3) cows losing condition at the time of mating 2 to 3 months after calving, which would affect the calf and the fertility if not supplemented, and (4) any cattle, store cattle or steers, which are almost marketable, but if not allowed to improve a little more might have to stay on the farm the best part of another year.

As mentioned earlier, supplementary feeding can be done by using deferred grazing systems, by a combination of grazing on pasture and feeding cut-and-carry fodder at night (Figure 9.5), or by using hay or silage produced earlier, if conditions allow for these practices.

Coconut Cake

It would be natural to use coconut press cake (called poonac in Sri Lanka) for supplementary feeding in areas where this material is available as a by-product from the oil extraction.

In Sri Lanka, feeding trials with Sinhala dairy cows (about 220 kg liveweight) over a two year period showed an increase in milk production per lactation of 60 per cent, when grazing was supplemented with 1.4 kg/day of coconut cake.

Treatments included 1.4 kg, 0.9 kg, and nil copra meal per cow per day. Milk production increased dramatically with supplementation; for example, cows fed 1.4 kg per day produced 420 liters per lactation, cows fed 0.9 kg produced 344 liters per lactation, and cows given no supplemental feed produced 260 liters per lactation. Lactation duration was increased through supplementation -- from 8 months for no supplementation, to 8½ and 9½ months for 0.9 and 1.4 kg coconut-oil meal, respectively. Also, cattle receiving coconut cake plus grazing developed a higher body weight and had more regular heat periods than those cattle which were only grazed.

Urea-Molasses

Non-protein nitrogen (NPN) in the form of urea has recently become popular to use as a source of nitrogen. The rationale for using NPN is that -- subject to a ready supply of energy being available

Figure 9.5.   A domestic buffalo herd feeding on
supplemental cut-and-carry fodder after being
grazed under coconuts during the day; Trinidad.

in the rumen -- the rumen bacteria will multiply, and convert NPN into bacterial protein. The increased microbial population thereby achieves better breakdown of the fibrous forage which is then eaten in greater quantity.

Urea is poisonous if taken in excess, and care must be given to assure that the animals cannot eat too much. One way to assure this is to dilute it with a molasses-water solution and dispense the mixture from one or more revolving drums, so arranged in the holding container that only a thin film can be ingested, and only then by the laborious process of licking. As well as being the diluent, the molasses acts as the readily-available energy source and renders the urea palatable (Yates and Schmidt, 1974). One roller drum container of 200 liter capacity will feed 50 head, and will need to be filled once a week. The mixture will be consumed at the rate of 20 kg urea and 50 liters of molasses in a 200 liter mixture (the rest being water) (Osborne, 1978, pers. comm.).

Other Feed Supplements

In the tropics farmers must be constantly alert to possible supplementary feeds from by-products of agro-industry and other sources. A common source of supplementary feeds is growth of naturalized grasses and shrubs along roadsides or in waste areas that can be cut and carried to the animals. Such plants include napier grass, guinea grass, para grass, leucaena (ipil ipil in the Philippines), and other herbaceous or edible shrubby plants. Common sources of by-product feeds include: (1) rice straw; (2) copra meal (coconut press cake); (3) rice hulls; (4) maize or sorghum stover; (5) sugarcane tops; (6) sweet potato vines; (7) cull fruits such as bananas, etc.; (8) unmarketed root crops such as sweet potato and cassava; (9) pineapple leaves, bran or peels; (10) crop residues from plants such as soybean, cowpea, and other pulses or field crops; (11) fish meal or other dried by-products, and many others. In Trinidad, Dr. Steve Bennett, a local veterinarian who is also a farmer, feeds poultry manure/litter (from sugarcane bagasse) to domestic buffalo up to 30 per cent of their total ration, with good results. It remains for farmers and agriculturists to find potential feed sources in their local communities and industries and to work out ways to make them useful in coconut/pasture/ livestock systems.

The College of Tropical Agriculture of the
University of Hawaii has been involved in research
on use of tropical products and by-products as livestock feed sources. Also, ASPAC has undertaken a
strong program to find ways to improve livestock
production through use of local feeds (Perez, 1976;
de Guzman, 1976).

Mineral Supplements

Mineral supplements usually contain calcium,
phosphorus, and salt. Calcium and phosphorus usually are provided in the form of dicalcium phosphate;
both are important constituents of bones and of soft
tissues.

Tropical pastures provide widely-varying mineral
intakes, and generalizations are best avoided. However, many extensive grazing areas of the tropics
are, in spite of the low growth rates they permit,
sufficiently lacking in these essential minerals to
allow clinical calcium/phosphorus deficiency to
appear in animals grazing on them. The ideal calcium
to phosphorus ratio in the feed is 1.5:1, but there
is considerable tolerance to variations from this,
especially when vitamin D levels are adequate.

Low phosphate levels are very common in the
tropics, due to high phosphate fixation in the soil.
This can be corrected by using dicalcium phosphate,
rock   phosphate, ground bone-meal, or -- in
selected cases or where calcium is in over-supply --
the addition of mono-sodium phosphate to the drink-
ing water, using a dispenser to provide a predetermined amount of mineral per liter of water. The
production of a pasture with a balanced mineral
composition is possible through fertilization, but
costs usually rule out such methods where grazing
areas are extensive. In most areas it is worthwhile
to provide ad-lib phosphate-rich licks or mixtures.

Other minerals might be required under some
conditions. Symptoms of copper deficiency are loss
of condition, loss of appetite, fading of the coat,
anemia, and suppression of oestrus. Forages from
affected areas usually have copper figures below
7.5 ppm (dry matter).

Coconuts are grown widely on coastal sands;
such soils may have copper or other mineral deficiencies. Copper and phosphorus deficiency has been
discovered in cattle in the Morobe district of Papua
New Guinea, a coconut-growing area (Mayall, 1973).
Similar symptoms as described for copper deficiency

can be caused by cobalt deficiency. Vitamin B$_{12}$, which is essential for haemoglobin formulation cannot be synthesized without cobalt, and a shortage of cobalt therefore causes anemia. Copper and cobalt can be provided through the provision of an appropriate mineral salt. Care must be taken to ensure that excessive amounts of copper cannot be taken, as the element can be toxic. Iodine deficiency or thyroid disfunctions have been reported in the tropics; this can be cured by providing a salt lick containing 0.02% potassium iodide.

HANDLING AND WORKING ANIMALS

Cattle should be "worked" (includes operations such as drenching, branding, dehorning, castrating, culling, selecting for market, spraying, dipping, weighing, checking for parasites and diseases, etc.) every so often so as to ensure that no problems exist. In some cases this may occur every 6 weeks to two months.

Sometimes docile cattle can be worked without special measures of restraint; however, for most beef animals this is usually not possible. Therefore, some sort of corral or holding pen will be needed to muster and restrain the animals so they can be examined at close quarters. For more active operations such as branding or castrating, a cattle squeeze (crush) may be required to prevent injury to the workers or the animal. Corrals, squeezes, and loading chutes do not have to be elaborate or expensive, but they should be sturdy and well-constructed in order to hold the animals. Figure 9.6 shows a corral, chute (race) and loading chute on a small coconut farm in Trinidad. This structure is somewhat light-weight for more active, large animals.

POISONOUS PLANTS

Many plants are known or suspected to be poisonous to livestock. It might be worthwhile to list and discuss briefly some of the plants likely to be found in coconut plantations that may injure cattle.

(1) Cassia occidentalis (coffee senna). A small leguminous shrub found in medium to high rainfall areas. The plant contains a laxative

261

Figure 9.6. A corral, chute (race), and loading chute on a coconut farm in Trinidad.

substance which may cause animals to become weak or even die, if sufficient quantities are eaten.

(2) Lantana camara (lantana). This weedy shrub is common in poorly managed coconut. It is suspected that it can cause photosensitization (makes animals become susceptible to severe sunburn), especially on white or unpigmented sections of the skin. The injury occurs because, after the animal eats lantana, damage is done to his liver.

(3) Xanthium strumarium, X. spinosum and others of the "cocklebur" group. Cocklebur is toxic in the very young seedling stage, especially when only the young "seed" leaves are present on the plant. Such seedlings are common around waterholes, shallow flood plains, or along streams and rivers. Animals may become prostrate and in severe cases may die.

(4) Ricinus communis (castor bean). The seeds, and to a lesser extent, the foliage of castor bean contain ricin, a very toxic poison. Livestock can die if sufficient quantities are eaten; however, poisoning occurs most frequently when animals eat infested feed grains.

(5) Sorghum halepense (Johnson grass). Johnson grass can be poisonous because of high concentrations of prussic acid (hydrocynaic acid) at certain times, especially following drought or cutting. At this time young shoots may contain much prussic acid. Nitrate poisoning can also occur from eating Johnson grass.

(6) Tribulus terristris and T. cistoides (puncturevine, caltrop, burhead). These prostrate, mat-forming vines with thorny, woody fruits are common along sandy beaches and waste places. Like lantana, they cause animals to become severely sunburned and sensitive to light because of liver damage after eating the plants.

(7) Leucaena leucocephala (leucaena, koa haole, ipil ipil). This shrubby legume which can grow to become a tree, can cause hair loss in horses and swine because of the presence of mimosine. It is a good pasture crop for tropical areas. In Australia and other places, if fed as the sole or major part of the ration, it can cause a goiter condition in cattle.

(8) <u>Argemone</u> <u>mexicana</u> (prickly poppy, mexican poppy). This plant has been suspected of livestock poisoning. Only hungry cattle will eat it. The active principle is an alkaloid.

10
Special Problems
and Considerations
for Small Farmers

Small farmers produce much of the world's coconut products. However, the small farmer suffers most from the problems of the industry, notably price fluctuations; injury or death of palms caused by disease, storms, or insects; and increasing production costs, especially costs of weed control.

Often, the farmer may have to choose between intercropping for food and pasture/livestock production systems. In many cases, however, the small farmer can combine intensive crop production under coconuts on his better lands with pasture/livestock systems on the poorer lands. To enable successful use of coconut lands for such mixed farming systems or, alternatively, to ensure success of pasture or fodder production by small farmers, some management steps or considerations especially tailored for small farmers may be necessary. This chapter will be devoted to a discussion of some of these considerations. Several of the preceding chapters also include factors affecting the small producer.

LAND TENURE

Land ownership or tenure may significantly affect small farmer production systems. For example many coconut lands are owned by communal or tribal groups, and any management system involving pasture/livestock production or intercropping with food crops must consider how best to satisfy the requirements of equitable sharing for the people concerned in tending and managing the communal lands as well as in distributing the food or income from the farm.

The Sili Village Project, Western Samoa

One example of a communal system which is meeting some of these issues is the Sili Village Development Program on the island of Savai'i, Western Samoa. This is not a government-sponsored program, but was devised by the "matai" or chief of the village, Leota Pita Ala'ilima, a former staff member of the University of the South Pacific School of Agriculture, Alafua, Western Samoa. The village owns a large quantity of land, much of which was poorly or little used except for individual family garden plots or occasional production of cash crops for short periods.

The village under the urging and leadership of Leota decided to develop its lands using village labor to clear and prepare land as well as to plant more than 400 ha of coconuts. While the young palms were growing and reaching bearing age, villagers grew taro, cassava and other food crops as intercrops with the coconuts. When palms reached four to six years of age and were large enough to escape livestock damage, cattle were brought in to graze the natural pastures (mostly sour paspalum, Paspalum conjugatum, and sensitive plant, Mimosa pudica) which developed in the understory. This provided weed control for coconuts as well as feed for the cattle. It is planned that when cattle numbers exceed the weed control requirements, the village will begin to plant improved pastures.

The village will obtain both milk and meat for local use. When production exceeds the local demand on the island, the cattle will be transported by interisland barges or ships to the market in Apia, the capital of the country.

Tanga, Tanzania

A very interesting project in Tanzania which overcame land tenure problems that prohibited grazing under coconuts was described by Childs and Groom (1964). Grazing was not possible because each farmer owned small, fragmented, scattered pockets of palms but only had usufructory rights to the land which he farmed. The small pockets of palms were too scattered to make concentrated food cropping or grazing of animals possible. Adding to the problem were the scattered, unfenced patches of food crops, mostly cassava, throughout the coconut lands.

266

Under these conditions care of the groves was inadequate, and undergrowth became rampant and out of control. To overcome the weed growth, annual fires were set to burn off the vegetation, resulting also in damage to palms. Under the prevailing conditions, cattle often had to walk long distances to reach water or grazeable forage. To add to the complications, **herdsmen** cared for the cattle, **most** of which belonged to many owners.

Previous work at the Tanga Livestock Experiment Station had shown that grazing dairy cattle under the palms controlled weed growth, improved milk production, and raised coconut yields. This work provided the technical basis for a project which sought to alleviate the land tenure and related management problems.

First of all, steps were taken to: (1) consolidate food-crop plots in a contiguous site near the villages, and (2) assist in obtaining grazing agreements between herdsmen (and livestock owners) and owners of the scattered coconut plots. Assistance was provided to the project by: (1) supply of herbicides to help control brush and tree weeds in the pastures, (2) provision of livestock water by supplying concrete rings for the lining of wells dug by communal effort in the sandy soil, (3) supply of barbed wire for perimeter fences as well as fencing of food crop areas, and (4) improved livestock management through loan of bulls for breeding, disease control and dipping, and improved methods of milk production.

A joint committee of livestock farmers and coconut farmers administers the program. The Ministry of Agriculture provides an extension advisor. The committee decides on plans for land clearing and management, communal grazing, areas to be used for food crops, and other related matters.

Yields of milk and coconut have increased significantly. Pastures have improved and are less weedy. Animal performance and management have improved markedly. The water supplies have been used for both domestic as well as livestock use and have improved the quality of life of the people.

The two case histories above illustrate several principles which can be of benefit to small farmers: (1) land consolidation to improve the management system and practices, (2) collaboration between herdsmen or livestock farmers with crop farmers to achieve common benefits, (3) use of traditional communal customs and practices to improve farm management as well as village welfare, and (4) supply of needed services and goods by government organizations or other groups.

Other solutions or methods are possible. The example of the system employed on Niue Island in the South Pacific (Chapter 1) is another example of overcoming land tenure problems or traditional farming patterns which hinder grazing or more intensive farming under coconuts.

MANAGEMENT PROBLEMS AND FACTORS

Managing the coconut/pasture/livestock system can complicate life for the small farmer. He may have to deal with new crops or livestock. Also, instead of caring for one crop, he now has two or three major components in his enterprise.

Lack of livestock experience may be one of the major limiting problems facing the small farmer. Some farmers may be afraid of cattle. Others may not recognize or understand the need for solution of such problems as animal health, breeding, herd management, or feeding needs. The problem of lack of experience need not be insurmountable, however.

Much can be done to overcome lack of experience with livestock by (1) starting with the right class or type of livestock; (2) starting with one or two docile animals such as dairy cattle; (3) education of farmers in basic animal health or management needs through special short courses, field visits, or pertinent publications; (4) devising simple systems which will fit the requirements of the small farmer, e.g. tethering draft or milk animals to palms, arranging for agreements between livestock owners needing pasture and coconut farmers, or starting the inexperienced farmer with animals which are less affected by low levels of management skills. A good example of the latter is a program in Papua New Guinea where small

farmers lacking livestock experience are provided
with steers that can be sold for cash after being
fattened on pasture (Jeffcott, Pers. Comm., 1972).
New steers are then purchased to replace those
sold. This system provides cattle experience for
the farmer without causing him to deal with the
breeding of cows, caring for young calves, and
segregation of animals according to age, sex, or
other criteria. As he gains experience with steers
he will later be able to take on herd management
problems with less difficulty.

Lack of experience in pasture management can
be a problem. However, a definite advantage of the
coconut/pasture/livestock system is that, until
grazing pressure to provide weed control exceeds
the available feed supply, there is little need to
plant improved pastures. In some cases, small
farmers may have so little land that they may wish
to increase forage production on their limited land
resources by planting improved pastures at the
start. In such cases a forage or forage mixture
should be chosen which will give satisfactory
yields under lower levels of management. Grasses
or legumes which can resist or survive overgrazing
or abuse should be used in such situations.

Farmers should be encouraged to think of their
pastures as a crop, and to manage them as well as
their resources or skills allow. The benefits will
be realized through improved livestock performance
as well as higher coconut yields.

Fencing

Livestock must be controlled, and fencing is
the best way to control loose animals. Fencing
costs can be high, even prohibitive in some cases.
Still, some solution can usually be found.

Stone fences can be useful in rocky areas
where some clearing of stones in fields is necessary. In such cases the fencing materials are
available and essentially free, except for the
labor involved in gathering and piling them. Rock
walls are widely used in the South Pacific islands
in coconut lands. Benefits are at least two-fold,
land improvement and control of stock.

Other locally-available materials for fences include poles or rails, bamboo, brush, etc. Sometimes "living fences" or hedges can be used. Gliricidia hedges have been recommended for use as living fences in Sri Lanka (Salgado, 1953). Forage shrubs such as Leucaena can be used as living posts while providing supplemental browse or cut forage. In the South Pacific kapok trees are used as living posts.

For some areas barbed or smooth wire may be the only materials available to provide for fencing. Often fencing wire must be imported and costs may be high.

Posts are another factor in fencing. Local trees are probably the least expensive source for posts. The life and durability of posts will vary widely according to type of tree used and soil type. Preservatives such as creosote should be used if possible in order to prolong the useful life of posts.

Steel or metal posts have the advantage of long life, but have a drawback in that they often must be imported and their cost is relatively high. It is possible to string fence wire from palm to palm, thus reducing the need for posts.

Cross fencing to subdivide pastures is important in rotational grazing systems. Construction of subdivision fencing can be scheduled over a period of time, thus reducing capital requirements in the early stages of pasture development.

An added benefit of fencing is prevention of damage to intercrops, pastures, or palms by marauding village livestock which are often unrestrained and roam at will through coconut lands (de Silva, 1953). Village pigs cause much crop and pasture damage in the South Pacific.

Fencing requirements may be reduced through tethering to palms, or penning and hauling cut feed to the animals in zero grazing systems. However, for most farms, at least the perimeters must be fenced to protect the crop and to control livestock.

Planting Materials and Seeds for Pasture Plants

Small farmers may have difficulty in obtaining planting materials or seeds of improved pasture species. The major commercial sources of tropical pasture grass and legume seeds are Australia, Brazil, and South Africa. Therefore seeds (and the appropriate Rhizobium strains for legumes) must be imported. Small farmers will not have the capability to import seeds; therefore commercial or government organizations will have to become involved in planning for importation, storage and distribution of seed-stocks. Small farmers will require only a few kilograms of most pasture seeds. Someone must help them to obtain needed supplies on a timely basis and at reasonable cost.

One way to reduce seed importation requirements is to plant nurseries of improved species, whether propagated vegetatively or from seeds, and use these nurseries to produce seed or planting materials for the surrounding farms. It may be that cooperatives or government agencies will need to supply this service.

If seed importations are required, the calibre of storage and handling management will be extremely important. Importations should be scheduled as closely as possible to coincide with the most suitable planting season. In the tropics most seeds will not remain viable for very long unless refrigerated, low humidity storage is provided.

Animal Management

Breeding stock. Most small farmers will not be able to afford keeping a bull for breeding purposes. Sharing a bull with a group of neighbors or borrowing a bull from government farms may be the best option. MacEvoy (1973) considers 15 ha as the minimum size farm on which beef breeding herds should be kept, and that a farmer should have at least 15 females if he is to keep a bull. If artificial insemination services are available, this may be the best solution for the small farmer, for it will increase conception rates and is less expensive. Borrowing bulls from government farms or superior farmers or use of artificial insemination will usually give the farmer access to superior breeding stock, thereby allowing an opportunity to upgrade his stock.

Controlling and handling bulls on small farms is difficult. Segregation of young heifers from the bull is just one of the complications.

Classes of cattle. As has been mentioned, steers may be the most suitable way for a farmer to get started in cattle production from coconut pastures. Docile milking cows may be next best in the system. Certainly many farmers will own draft or domestic milk animals which can be grazed under the palms. Dual-purpose animals which can provide both milk or meat may be most suitable for many small farms.

Breeds of cattle. Are there certain breeds which may be more suitable for small farmers? This is a difficult question. Certainly cattle belonging to the genus Bos indicus usually do perform better under tropical conditions. However, the cool shady environment of pastures under coconuts does provide a more favorable situation for European breeds than the more open, unshaded lands.

Because many coconut pastures are used primarily for milk production, dairy breeds are used on many small farms.

In Sri Lanka much effort has been devoted by the Coconut Research Institute to improving the small local Sinhala cattle for milk production under coconuts. A Jersey-Sinhala cross has proved to be more productive than the Sinhala breed. The program is specifically aimed at improving milk production on small coconut farms. For details see Goonesekera (1967); Appadurai (1968); and Buvanendran and Mahadevan (1975).

Other classes of livestock. Cattle or buffalo are certainly not the only animals which can be raised under coconuts. In Sri Lanka there has been considerable interest in raising sheep in coconut lands. This has proved to be a very satisfactory system (Appadurai, 1968; Perera, 1972; van der Porten, 1972). Sheep are easily fenced in and restrained, but they have the disadvantage of being very susceptible to pests and diseases in tropical areas. Wool quality and the need for shearing can complicate the management system for small farmers.

There has been recent interest in using sheep for milk production (Gall, 1975).

Goats are often grazed under coconuts, but usually are unrestrained and therefore contribute to overgrazing and resultant soil and pasture degradation. Goats are sometimes tethered to palms in Sri Lanka. With proper fencing and management, goats could provide both meat and milk in coconut areas.

Soil compaction by animals may be a problem on small farms where tethering to palms is practiced or when heavy concentrations of livestock on pastures during wet periods may occur. Compaction problems will be reduced if a vigorous forage crop covers the soil.

Manure or compost for crop production can be a real benefit of animal production on small farms (Shrader, 1951; de Silva, 1961; Anonymous, 1967). Penning or housing at night will ensure that manure can be collected and used to fertilize palms or food crops. Penning or confinement will also decrease the problems of rhinoceros beetle buildup in dung in coconut/pastures and the effect of urine and dung in producing lush but less palatable forage.

Zero grazing (cutting and hauling forage to penned livestock) may be a very suitable option for small farmers. Advantages are: (1) forage or soilage crop production can be intensified on certain land areas; (2) superior, high-producing forages which are unsuited or poorly suited for grazing can be used to best advantage (napier grass and Leucaena are examples of plants which can be used in this way);(3) forage utilization is improved because trampling and fouling of forages under grazing are reduced, additionally, forage losses because of selective grazing are reduced; and (4) fencing and animal control problems are reduced. Better labor utilization may also be a benefit.

Zero grazing or cut forage systems can be combined with grazing of seasonal pastures, feeding of crop residues, or supplementation on pastures to provide a viable feed system.

273

Access to credit for small farmers may become
a problem. Capital will be required to purchase
fertilizers, breeding stock, feeder animals, seed,
fencing, or other inputs. Governments which recognize the potential economic and food production
benefits of coconut/pasture/livestock systems can
accelerate development by ensuring reliable credit
under reasonable terms for the small producer.

What is an economic unit? This is an open
question, depending on the farmer and his personal
situation. Family food or income needs, land and
labor resources, age of palms, and access to markets can affect the scale and type of farming
enterprise which he employs. The weed control
benefits of grazing may be the first consideration
in determining livestock numbers, marketable or con-
sumable products, or the coconut/pasture system.
MacEvoy (1973) suggested that farms 15 ha or more
in size could be considered as large enough for
beef breeding operations, while those 15 ha or
less should be used for fattening store animals.

For detailed discussions of the economics of
pastures under coconut, the careful farm management
studies of Barker and Nyberg (1968) and de Guzman
(1970) in the Philippines, Antoine (1973) in
Trinidad and Tobago, and Selvadurai (1968) in
Malaysia are commended to the readers.

11
Research, Research Needs, and Future Outlook

An awareness of the importance of coconut/ cattle enterprises is growing in many countries and will probably continue. Examples of the growing interest are available in the South Pacific Islands and in Southeast Asia. Research and development programs in coconut/pasture systems, once restricted mostly to Sri Lanka, have been initiated in the Solomon Islands, Fiji, India, Indonesia (Bali), Jamaica, Malaysia, Niue, Papua New Guinea, the Philippines, Tanzania, Thailand and Western Samoa.

International meetings and workshops on coconut/pasture systems have been held recently in Western Samoa, the Philippines, and the Solomon Islands. These meetings have helped to dramatize the farming system and to mobilize interest in extending it to other areas.

South Pacific Regional Seminar on "Pastures and cattle under coconuts"

The first international meeting devoted to coconut/pasture/beef was sponsored by the South Pacific Commission and was held at Alafua College in Apia, Western Samoa on August 30 to September 13, 1972. The guiding force behind the meeting was Mr. Edwin I. Hugh, then Livestock Specialist for the South Pacific Commission. He conceived the need for such a seminar and led in developing and organizing it. Officially called the "South Pacific Regional Seminar on Pastures and Cattle under Coconuts," the seminar brought together a group of agriculturists with interests in both pasture and animal production from the Pacific Islands and countries including Fiji, New Hebrides, American

Samoa, Solomon Islands, Niue, Tonga, Western Samoa, New Caledonia. Australia, Papua New Guinea, and USA (Hawaii). The United Nations Development Program in Western Samoa was also represented.

Agriculturists from each country or territory represented brought information on the current status of coconut/pasture/cattle systems in their home jurisdictions.

Consultants and resource persons in pasture management, animal management, and intercropping with coconut presented background and resource papers for the meeting (Osborne 1972, Plucknett 1972, Robinson 1972). As individual topics were introduced, each country representative discussed its importance in his country. Working sessions and discussions were recorded and were summarized for inclusion in the final reports. A list of participants is presented in Appendix A.

A report of the seminar has been published by the South Pacific Commission, P.O. Box D 5, Noumea, New Caledonia (Hugh 1972; South Pacific Commission, 1972).

Research needs and recommendations suggested by the seminar are presented later in this chapter.

ASPAC Training course on "Pasture production under coconuts"

On October 15-19, 1973 the Food and Fertilizer Technology Center (FFTC) of the Asian and Pacific Council (ASPAC), headquartered in Taipei, Taiwan, Republic of China, held a training course on "Pasture Production Under Coconuts" in Davao, Mindanao, Philippines. The course was attended by scientists and farmers from Asian countries. Cosponsors for the training course were the FFTC and the National Food and Agriculture Council of the Philippines, with the cooperation of a number of national Philippine agencies.

Several publications have been issued by FFTC as a result of this training course (de Guzman, 1974; Javier, 1974; McEvoy, 1974). These bulletins have stimulated interest and activity in pasture/coconut management, and it is certain that new research programs will also result. Mr. Moises R. de Guzman, Jr. of the FFTC staff has had considerable experience with pastures under coconuts in the Philippines, and is the senior author of a very useful small book on the coconut/pasture/cattle system (de Guzman and Allo, 1975). A list of papers and resource papers is presented in Appendix B.

Regional Seminar on Pasture Research and Development in the Solomon Islands and Pacific Region.

On August 29 to September 6, 1977, the Australian Development Assistance Bureau, Government of Solomon Islands and the University of Queensland, Australia, sponsored a regional seminar on pasture research at Honiara, Solomon Islands. Dr. P. C. Whiteman of the University of Queensland was chairman of the organizing Committee. Representatives of most island nations and territories of Oceania participated, as did representatives of several developed countries. The seminar included technical papers and field trips demonstrating coconut/pasture research as well as current farming practices. A report of the seminar has been published (Ministry of Agriculture and Lands, Solomon Islands 1977). A list of papers presented is presented in Appendix C.

RESEARCH AND DEVELOPMENT PROGRAMS IN PROGRESS

ASIA ⁻

Sri Lanka

The Coconut Research Institute (CRI) started its coconut/pasture research program in the early 1950s. The program has emphasized inter-specific competition between coconuts and pastures, comparison of pasture species, fertilizer needs of pasture grasses and coconuts when grown in combination, subsidiary crop production, and cattle (mostly dairy) management in coconut/pasture systems. Their pioneering research confirmed the value of Brachiaria miliiformis (corigrass) as a pasture plant under coconuts. Both agrostologists and livestock specialists are involved in the program.

Today the CRI research on pastures and cattle under coconuts and intercropping with food and cash crops provides a sound basis for government development efforts in the Coconut Triangle. CRI is now conducting regional trials on these systems.

Intercropping trials are being conducted on small farms to encourage farmers to adopt the most successful practices. Government subsidies are provided to farmers who establish pastures under coconuts. The subsidy pays 90 per cent of the establishment cost.

A recent joint report by the National Science Council of Sri Lanka and the National Academy of Sciences of the USA (NAS, 1976) recommended that coconut intercropping and pasture systems be given high priority.

Leaders in the CRI research program over the years (and therefore pioneers) were G. Goonasekera and Dr. K. Santhirasegaram. Douglas E.F. Ferninandez is now in charge of the pasture management program. Dr. U. Pethiyagoda is Director of the Institute.

The address of the Coconut Research Institute is:

Bandirippuwa Estate
Lunawila, Sri Lanka

Philippines

The College of Agriculture of the University of the Philippines at Los Banos (UPLB) has worked with various aspects of the coconut/pasture/cattle system, which is called "coco-beef" in that country. Research on improved pasture species and in cattle management is in progress. A newly-formed coconut industry group, the Philippine Coconut Authority (PCA) should help to stimulate further interest in the system. The PCA replaced the former Philippine Coconut Research Institute (PHILCORIN). The PCA proposes to conduct research on: (1) improved management of pastures and grasslands under coconuts and to raise production of existing coconut/cattle enterprises, (2) profitability of coconut/cattle enterprises, and (3) effect of ecological factors on coconut production (PCA Annual Report, 1973-74).

Many commercial farms and plantations on the island of Mindanao have practiced "coco-beef" production for years. The Philippine Council for Agricultural and Resources Research (PCARR) recently published a bulletin entitled, "The Philippines Recommends for Coconuts 1975."

278

Persons with experience in the system include the following professors at UPLB: Dr. Emil Q. Javier, Dept. of Agronomy; Dr. Joseph C. Madamba, formerly with the Department of Animal Sciences, and former Director of the Philippine Council of Agricultural and Resources Research; Dr. Juan T. Carlos, Department of Horticulture; Dr. Bonifacio C. Felizardo, Department of Soil Science; and Dr. P. Sajise, Department of Botany. Mr. Michael MacEvoy a member of the World Bank staff in Manila, is an expert on practical management problems of coco-beef on commercial farms. Dr. Pablo Manolo of the Bureau of Animal Industry is experienced in live-stock disease and pest problems under coconuts. Mr. Severino S. Magat is in charge of PCA research activities planned by the Agronomy-Soils Division, Agricultural Research Department, of the Philippine Coconut Authority.

Addresses:

Agricultural Research Department
Philippine Coconut Authority (PCA)
P.O. Box 295
Davao City, Philippines

College of Agriculture
University of Philippine at Los Banos
Los Bano, Philippines

Bureau of Animal Industry
Manila, Philippines

India

Indian farmers have practiced intercropping for some time. Research on intercropping and pasture or fodder production is conducted at the Central Plantation Crops Research Institute (CPCRI) at Kasaragod.

Persons knowledgeable about pastures and fodders under coconuts are K. N. Sahasranaman and K.S. Menon. Intercropping research is being con-ducted by E.V. Nelliat, Head of the Division of Agronomy of CPCRI and Dr. P.K.R. Nair, agronomist. K.V. Ahamad Bavappa is Director of the Institute.

The address is:

Central Plantation Crops Research Institute
Kasaragod, South India.

Thailand

A joint program to develop pasture and cattle
under coconuts has been initiated by the FFTC and
the Applied Scientific Research Corporation and
Livestock and Agriculture Departments of Thailand
(de Guzman, 1974; Allo, 1976). The program is
designed to demonstrate to farmers the potential
benefits and management practices of cattle under
coconut, to obtain information on best pasture
species to be used under coconuts, and the effects
of pastures and livestock on coconut yields. The
program is aimed at coconut farmers in southern
Thailand where there are about 100,00 ka of coconuts. Farmers are already grazing natural pastures
in this area; some have established improved pastures.

OCEANIA

South Pacific Commission

The SPC was the first international organization to recognize and promote the coconut/pasture/
cattle farming system (South Pacific Commission,
1972). Edwin I. Hugh, then Livestock Officer, was
the driving force behind the program.

In 1974, the 14th South Pacific Conference
agreed to give top priority in its Economic
Development Program to production of cattle under
coconuts (Anon. 1974).

Address: Post Box D5
           Noumea, New Caledonia

Western Samoa

A leader in practical management of cattle
and pastures under coconut is the Western Samoa
Trust Estates Corporation (WSTEC), a corporation
owned by the government of Western Samoa and
managed on a commercial basis in the public interest. WSTEC has experimented over the years in
making this farming system work on a commercial

basis, and many of their innovations have been adopted elsewhere. Their address is: Apia, Western Samoa. Mr. Morris Lee, Associate General Manager, and Willie Wong, Assistant General Manager, are experienced in the coconut/pasture/cattle system.

UNDP is conducting research on pasture under coconuts at Vailele, a WSTEC plantation. Cooperating in the research are UNDP, the Department of Agriculture of Western Samoa, WSTEC, and Alafua Agriculcultural College (The University of the South Pacific College of Agriculture). These trials were established to study production of natural and improved pastures, to find suitable legumes and grasses and their combinations for use under coconuts, and to determine the influence of management practices such as fertilizer use, grazing intensity, and pasture establishment. A number of reports have been released on this work (Ericksen, 1973; Reynolds, 1975, 1976a, 1976b, 1977a, 1977b, 1977c, 1977d, 1977e, 1977f, 1977g, 1978a, 1978b; Reynolds and Ericksen, 1976; Reynolds and Lovang, 1977; Reynolds, et. al. 1978; Reynolds and Schleicher, 1975; Reynolds and Uati, 1976).

Addresses of cooperators are as follows:

UNDP-FAO Special Project
c/o Dept. of Agriculture
Apia, Western Samoa

Director
Dept. of Agriculture
Apia, Western Samoa

Principal
University of the South Pacific **School of Agriculture**
Alafua, Western Samoa

Solomon Islands

A joint research program on coconuts has been conducted by the Ministry of Agriculture and Lands and Lever's Pacific Plantations Pty. Ltd. (Solomon Islands Joint Coconut Research Scheme, 1966-67, 1967-68, 1969). Most of this research is directed toward coconut directly, but some spacing and other management studies (Friend, 1977) do bear on management of natural understory pastures.

Addresses:

Ministry of Agriculture and Lands
P.O. Box 25
Honiara, Solomon Islands

Lever's Pty Ltd.
Yandina, Solomon Islands

Papua New Guinea

Research on pasture production and intercropping has been conducted at the Lowlands Agricultural Experiment Station on the island of New Britain (Bourke, 1976, 1978). Cacao is the main intercrop in coconut, and much research has been done on this combined perennial tree crop system.

Efforts are being made to develop pastures under coconuts and to increase cattle production.

Address: Lowlands Agricultural Experiment
          Station
          Keravat, East New Britain
          Papua, New Guinea

Fiji

The Department of Agriculture is conducting pasture research, with special emphasis being given to screening pasture legumes and grasses for coconut lands in the different ecological zones of the islands. The pasture agronomist is Eminoni Ranacou. James Makasiale is knowledgeable about traditional production systems of small farmers.

Address: Department of Agriculture
          Suva, Fiji

Niue

Some of the more critical research on the coconut/pasture system has been conducted on Niue Island by Richard Lucas of New Zealand on secondment to the Department of Agriculture in Niue (Lucas, 1968). Mr. Lucas is now on the staff of Lincoln College in New Zealand. The purpose of the research has been to establish improved pastures under coconuts for small farmers on the island. Competition studies and fertilizer requirements, including study of zinc deficiency on soils derived from coral, have been conducted.

> Address: Department of Agriculture
> Niue Island
> South Pacific

WEST INDIES

Jamaica

The Coconut Industry Board has conducted research on intercropping and pastures under coconuts, including experiments on performance of natural and improved pastures (pangolagrass, guineagrass and tropical kudzu) and their effects on coconut yield and nutrition (Coconut Industry Board, Jamaica, 1961, 1962, 1963, 1964, 1965, 1966, 1967, 1968, 1970, 1971). The Manager/Secretary of the Board is R.A. Williams.

> Address: Coconut Industry Board
> P.O. Box 204
> Kingston 10, Jamaica

Trinidad and Tobago

A very thorough and interesting research program on economic aspects of the coconut/pasture/beef cattle system on Trinidad and Tobago was conducted by Antoine (1973). He defined coconut/beef farms as those where the livestock population was at least one animal unit to every 1.6 ha of coconut land. Antoine reported the coconut area of Trinidad and Tobago at about 14,400 ha in 1963.

AFRICA

Tanzania

Research on managing pastures under coconuts for dairy production has been conducted at the Tanga Livestock Breeding Station. Studies on weed control, fertilization of palms, and insect control, coupled with an imaginative advisory service to bring about land reform, grazing agreements, fencing, and provision of livestock water have brought about increases in milk as well as copra production (Childs and Groom, 1964).

G.D. Anderson (Anderson, 1967) of the Northern Research Centre, Ministry of Agriculture, Forests and Wildlife, Tengeru, Tanzania has led the Tanga research program.

The address for the Tanga Station is:

Tanga Livestock Breeding Station
Tanga, Tanzania

DEVELOPED COUNTRIES

Australia

The Department of Agriculture of the University of Queensland has studied pastures under coconuts in the South Pacific (Whiteman, 1977; Gutteridge and Whiteman, 1977, 1978; Gutteridge, et al. 1976) and in Indonesia (Steel, 1974; Steel and Humphreys, 1974). Performance and management of grasses and legumes under coconuts have been emphasized. Leaders of the research are Dr. L. Ross Humphreys and Dr. Peter Whiteman. Their address is:

Department of Agriculture
University of Queensland
St. Lucia, Queensland, Australia

Dr. George Osborne, a staff member of the Veterinary School, has had experience with cattle management under coconuts as a consultant for SPC and private interests in the Solomon Islands and New Hebrides (Osborne, 1972, 1978). His address: Veterinary School, University of Queensland, St. Lucia, Queensland, Australia.

Hawaii (USA)

Research and state-of-art studies have been conducted on management of improved tropical forage species for coconut lands, especially in relation to shade tolerance of pasture species, weed and brush control, fertilizer needs, and performance of weeds and naturalized plants in unimproved pastures under coconuts (Plucknett, 1972a, 1972b, 1978; Plucknett and Nicholls, 1972; Ericksen, 1977; Ericksen and Whitney, 1977). Principal scientists are Dr. Donald L. Plucknett (management, weed control, land use, pasture ecology) and

Dr. A. S. Whitney (pasture physiology, tropical legume management). Dr. Flemming Erickson, a former student, (formerly with UNDP in Western Samoa and now residing in Denmark) specialized in tropical forage evaluation, comparisons of natural and improved pastures, and shade tolerance of pasture plants. Address:

Department of Agronomy and Soil Science
College of Tropical Agriculture
University of Hawaii
Honolulu, Hawaii, USA 96822

RESEARCH NEEDS

Participants at the South Pacific Commission Seminar on "Pastures and Cattle under Coconuts" identified a number of research topics and priorities for the coconut/pasture/cattle system in the South Pacific. Recommendations fell into five general categories: (1) effect of intercropping and grazing on coconuts, (2) natural pastures under coconuts, (3) improved pastures under coconuts, (4) animal management in the coconut/pasture system, and (5) economics and marketing. (It should be mentioned here that it was concluded State-of-Art studies should precede such research.)

A. Intercropping or grazing effects on coconut.

1. Soil compaction by cattle.

2. Tillage and cultivation; favorable or unfavorable? Frequency, depth, and time of cultivation.

3. Intercrop or pasture competition -- light, shade, etc.

4. Fertilization practice, -- time, rates, frequency, requirements for palms of different age.

5. Spacing of newly-planted palms for intercropping.

B. Natural Pastures

1. Need to survey and understand composition and yield of natural pastures.

285

2. Weed control, from uncontrolled under-story to managed natural pastures.

3. Natural pasture performance under different management regimes and levels of utilization.

4. Carrying capacity and animal productivity; optimal stocking rates.

5. Seasonal forage yields; obtaining a balanced feed year.

6. Forage value of natural plants.

C. Improved pastures

1. Establishment methods.

2. Fertilizer requirements for pasture grasses and legumes as well as coconut.

3. Weed control.

4. Finding suitable grass/legume mixtures which are shade tolerant and productive, but which do not depress coconut yields.

5. Carrying capacity, animal production, and optimum stocking rates.

D. Animal Management

1. Suitability of small versus large ruminants.

2. Cattle breeds and productivity.

3. Breeding systems.

4. Animal health and disease control.

5. Value of copra as animal feed (during periods of low prices).

E. Economics and marketing

1. Planning for improved and diversified land use under coconuts.

2. Farm size, income.

3. Marketing methods, organization.

4. Technical advice and assistance for farmers.

In carrying the recommended research a <u>general outline of steps to be followed</u> included:

1. Record and circularize a list of present work being done,

2. Register needs in order of priority,

3. Plan for dissemination and organization of information,

4. Recommend regional research projects and consider countries to handle each of these,

5. Plan regular meetings to discuss results and to plan for utilization of findings.

Solomon Islands Regional Seminar on Pasture Research

This seminar identified a number of research priorities for the coconut/pasture system (Solomon Islands Regional Seminar Pasture Research, 1977). A number of research priorities identified at the SPC seminar were reiterated at Honiara; however, topics receiving special attention at Honiara included the following: (1) introduction and management of suitable (shade tolerant, persistent, less competitive) pasture species and mixtures, (2) low input systems, (4) effect of particular grasses on coconuts, (5) plant nutrition, (6) local phosphorus fertilizer sources, and (7) management systems.

References

ABEYWARDENA, V. 1954. The density of palms in triangular and square planting. Ceylon Coconut Quarterly 52(2):94.

ABEYWARDENA, V. 1955. Rainfall and crops. Ceylon Coconut Quarterly 6(1 & 2):17-21.

ABEYWARDENA, V. 1962. The raingauge and estate management. Ceylon Coconut Planters' Review 2(4):3-13.

ABEYWARDENA, V. 1968. Forecasting coconut crops using rainfall data - a preliminary study. Ceylon Coconut Quarterly 19:161-176.

ABEYWARDENA, V. 1971. Yield variations in coconut. Ceylon Coconut Quarterly 22(3 & 4):97-103.

ABEYWARDENA, V. and J. FERNANDO. 1963. Seasonal variation of coconut crops. Ceylon Coconut Quarterly 14(3 & 4):78-88.

ACLAND, J. D. 1971. East African Crops. Food and Agriculture Organization of the United Nations. Longmans, London. 252 pp.

AGCOILI, L. B. 1973. Soil management for coconut farms. ASPAC Food and Fertilizer Technology Center Extension Bulletin No.35. Taipei, Taiwan.

AGENCY FOR INTERNATIONAL DEVELOPMENT. 1975a. Characteristics of economically important food and forage legumes and forage grasses for the tropics and sub-tropics. Technical Services Bulletin No.14, Office of Agriculture, Technical Assistance Bureau, Agency for International Development, Washington, D.C. 20523. 29 pp.

AGENCY FOR INTERNATIONAL DEVELOPMENT. 1975b. The contribution of legumes to continuously productive agricultural systems for the tropics and sub-tropics. Technical Series Bulletin No.12, Office of Agriculture, Technical Assistance Bureau, Agency for International Development, Washington, D.C. 20523. 42 pp.

AGENCY FOR INTERNATIONAL DEVELOPMENT. 1976. Dominican Republic Fertilizer Situation; A Description, Analysis and Recommendation. Bulletin Y-103, National Fertilizer Development Center, Tennessee Valley Authority, Muscle Shoals, Alabama. 101 pp.

ALBUQUERQUE, S.D.S. 1964. Coconut cultivation in Mysore. Coconut Bulletin 18(6):249-256.

ALFEREZ, A. C. 1975. Coconut/cattle farming: problems and potentials. National Coconut Research Symposium, Tacloban City, November 17 - 19, 1975. Philippine Council for Agriculture and Resources Research. p. 170-178.

ALLO, A. V. 1976. Pasture - coconut project ASPAC Food and Fertilizer Technology Center Newsletter, March, 1976. 2 pp.

AMBROSE, C. 1952. The root system of the coconut palm. Ceylon Coconut Quarterly 3(1):16-20.

ANDERSON, G. D. 1967. Increasing coconut yields and income on the sandy soils of the Tanganyika coast. East African Agricultural and Forestry Journal 32(3):310-314.

ANONYMOUS. 1929. Green manuring with special reference to coconuts. Tropical Agriculturist, Ceylon. 73(3):144-155.

ANONYMOUS. 1936. Use of Leguminous Plants in Tropical Countries as green manure, as cover and as shade. International Institute of Agriculture, Rome.

ANONYMOUS. 1951. Report of the technical committee on marginal lands. Ceylon Coconut Quarterly 2(2):138.

ANONYMOUS. 1952. Report, Division of Animal Husbandry. Ceylon Coconut Quarterly 3(1):43-44.

ANONYMOUS. 1953. Cover crops suitable for coconut estates. Ceylon Coconut Research Institute. Leaflet No.3.

ANONYMOUS. 1954. Report, Division of Animal Husbandry. Ceylon Coconut Quarterly 5(1):29-31.

ANONYMOUS. 1958. Control of illuk. Coconut Research Institute of Ceylon Leaflet No.28. 4 pp.

ANONYMOUS. 1960. The manuring of adult coconut palms. Ceylon Coconut Research Institute Leaflet No.36. 7 pp.

ANONYMOUS. 1962. Brachiaria brizantha - the perfect pasture. Ceylon Coconut Planters Review. 3(1):22-25.

ANONYMOUS. 1964. Intercropping of coconut gardens in Andhra Pradesh. Coconut Bulletin 18(8):343-345.

ANONYMOUS. 1966a. In Andhra Pradesh banana is a popular intercrop in coconut gardens. Coconut Bulletin 19(10 & 11):306-308.

ANONYMOUS. 1966b. Pasture under coconuts. Ceylon Coconut Research Institute Advisory Leaflet No.45.

ANONYMOUS. 1967a. Cattle under coconuts -- farmyard manure. Ceylon Coconut Research Institute Advisory Leaflet No.24.

ANONYMOUS. 1967b. Manuring of young palms. Ceylon Coconut Research Institute Leaflet No.8. 5 pp.

ANONYMOUS. 1967c. The use of locally available organic materials for manuring coconuts. Ceylon Coconut Research Institute Leaflet No.9. 10 pp.

ANONYMOUS. 1971a. Conservation of soil and moisture. Coconut Bulletin 2(3):7-10.

ANONYMOUS. 1971b. Profitable subsidiary crops. Coconut Bulletin 2(1):6-7.

ANONYMOUS. 1972. An improved method of cultivation of Crotalaria striata. Coconut Bulletin 2(9):10-11.

ANONYMOUS. 1973. Intercrop while coconuts are young. Animal Husbandry and Agriculture Journal 8(6):14.

ANONYMOUS. 1973-74. Annual Report, Agricultural Research Department, Philippine Coconut Authority. P.O. Box 295, Davao City, Philippines.

ANONYMOUS. 1974. Report of the Fourteenth South Pacific Conference. South Pacific Quarterly 24(4):17.

ANONYMOUS. Undated. Coconut Production in the Philippines. College of Agriculture, University of the Philippines at Los Banos. 51 pp.

ANTOINE, A. 1973. The Economic and Financial Implications of Introducing Beef Cattle on Coconut Farms in Trinidad and Tobago. unpublished Master of Science thesis, University of the West Indies, St. Augustine, Trinidad and Tobago. 135 pp.

APPADURAI, R. R. 1968. Grassland Farming in Ceylon. T.B.S. Godamunne & Sons Ltd., Kandy, Sri Lanka. 135 pp.

ASIAN DEVELOPMENT BAND (ADB). 1969. Asian Agricultural Survey. University of Tokyo Press, Tokyo, Japan. 787 pp.

BALAKRISHNAMURTI, T. S. 1969 Isotope studies on efficiency of fertilizer utilization by coconut palms. Ceylon Coconut Quarterly 20:111-122.

BALASUNDARAM, E. K. AND S. G. AIYADURAI. 1963. Grow spice crops in the shade of coconut palms. Coconut Bulletin 17(5):182-184.

BARKER, R. AND A. J. NYBERG. 1968. Coconut-cattle enterprises in the Philippines. Philippine Agriculturist 52(1):49-60.

BEDFORD, G. O. 1972. Rhinoceros beetle. In: Coconut Industry Workshop. Department of Agriculture, Fiji. p. 45-51.

BOOLINKAJORN, P. AND DURIYAPRAPAN, S. 1977. Herbage yield of selected grasses grown under coconuts in southern Thailand. Thailand Journal of Agricultural Science. 10:35-40.

BOURKE, R. M. 1975. Evaluation of leguminous cover crops at Keravat, New Britain. Papua New Guinea Agricultural Journal 26(1):1-9.

BOURKE, R. M. 1976. Food crop farming systems used on Gazelle Peninsula of New Britain. Proceedings, Papua New Guinea Food Crops Conference, 1975. Department of Primary Industry, Port Moresby, Papua New Guinea. p. 81-100.

BOURKE, R. M. 1978. Suggested farming systems for lowland forest areas where land is scarce. Harvest (Papua New Guinea). 4(3):179-187.

BOYENS, R. W. 1972. Pasture establishment under coconut palms. Regional Seminar on Pastures and Cattle Under Coconuts. Report of meeting, Alafua, Western Samoa, August - September 1972. South Pacific Commission, Noumea, New Caledonia. p. 105-108.

BUCHANAN, D. W. 1966. Coconut nutrition. In: Fruit Nutrition (N. F. Childers, ed.). 2nd ed. Horticultural Publications, Rutgers, New Jersey. Chapter XIX. p. 597-610.

BUNTING, B. 1926. Observations of cover crops at Castleton Estate. Tropical Agriculturist, Ceylon 66(1):43-47.

BURCHAM, L. T. 1947. Livestock grazing in the Russell Islands. Journal of Forestry 45:113-117.

BURGOS, C. H. 1938. You can raise livestock in coconut groves. Agricultural Industry Monthly 5(11):12-13, 36, 37, 38, 44.

BUVANENDRAN, V. 1970. Breeding for beef. In: The Development of The Cattle Industry in Ceylon. (R. R. Appadurai, Ed.). Proceedings of a Symposium, University of Ceylon and The Ministry for Agriculture and Food, Colombo, Sri Lanka. p. 131-139.

BUVANENDRAN, V. AND P. MAHADEVAN. 1975. Cross-breeding for milk production in Sri Lanka. World Animal Review (FAO) 15:7-13.

CALO, LITO L. (Undated). Cattle and Carabao: priority for development in Southeast Asia. Unpublished Manuscript, The Food Institute, East West Center, Honolulu, Hawaii 96822. 34 pp. (typewritten).

CARRAD, B. 1977. Cattle and Coconuts: A Study of Copra Estates in the Solomon Islands. A draft report for the South Pacific Commission. Australian National University, Canberra, Australia. 131 pp.

CATLEY, A. 1969. The coconut rhinocerous beetle Oryctes rhinoceros (L.). PANS 15(1):18-30.

CELINO, M. S. 1963. Increasing income from coconut lands. Coconut Bulletin 17(5):169-172.

CELINO, M. S. 1964. Coconut industry in the Philippines. Oleagineux 19(1):19-25.

CHALMERS, A. 1968. Establishing cocoa under coconuts: the early stages. In: Cocoa and Coconuts in Malaysia. J. S. Blencowe and P. D. Turner (ed.). Proceedings of a Symposium in 1967. The Incorporated Society of Planters, P.O. Box No.262, Kuala Lumpur, Malaysia. p. 12-19.

CHARLES, A. E. 1959. Coconut agronomy 1954 -1958. Papua and New Guinea Agricultural Journal 12(1): 28-36.

CHARLES, A. E. 1964. Coconuts on coral-derived soils respond to potassium. Coconut Bulletin 18(5):219-222.

CHEYNE, O. B. M. 1952. Revised estimate of expenditure on replanting (underplanting 1 acre of coconuts). Ceylon Coconut Quarterly 3(3): 133-134.

CHILD, R. 1955. The coconut industry in Portuguese East Africa. World Crops 7:488-492.

CHILD, R. 1964. Coconuts. Longman, London. 216pp.

CHILDS, A. H. B. AND C. G. GROOM. 1964. Balanced farming with cattle and coconuts. East African Agricultural and Forestry Journal 29(3):206-207.

COCKRILL, W. R. (ed.). 1974. The Husbandry and Health of the Domestic Buffalo. Food and Agriculture Organization of the United Nations, Rome. 993 pp.

COCONUT INDUSTRY BOARD, JAMAICA. 1961. Annual Report of Research Department, April 1959 - June 1961. Coconut Industry Board, Kingston, Jamaica. p. 10.

COCONUT INDUSTRY BOARD, JAMAICA. 1962. Annual Report of Research Department, 1961 - 1962. Coconut Industry Board, Kingston, Jamaica. p.16.

COCONUT INDUSTRY BOARD, JAMAICA. 1963. Annual Report of Research Department, 1962 - 1963. Coconut Industry Board, Kingston, Jamaica. p.21.

COCONUT INDUSTRY BOARD, JAMAICA. 1964. Annual Report of Research Department, June 1963 - June 1964. Coconut Industry Board, Kingston, Jamaica. p. 24-25.

COCONUT INDUSTRY BOARD, JAMAICA. 1965. Annual Report of Research Division, 1964 - 1965. Coconut Industry Board, Kingston, Jamaica. p.24.

COCONUT INDUSTRY BOARD, JAMAICA. 1966. Annual Report of Research Division, 1965 - 1966. Coconut Industry Board, Kingston, Jamaica. p. 50-51.

COCONUT INDUSTRY BOARD, JAMAICA. 1967. Annual Report of Research Division, 1966 - 1967. Coconut Industry Board, Kingston, Jamaica. p. 63-65.

COCONUT INDUSTRY BOARD, JAMAICA. 1968. Annual Report of Research Division, 1967 - 1968. Coconut Industry Board, Kingston, Jamaica. p. 49-50.

COCONUT INDUSTRY BOARD, JAMAICA. 1970. Annual Report of Research Division, 1969 - 1970. Coconut Industry Board, Kingston, Jamaica. p. 37-41.

COCONUT INDUSTRY BOARD, JAMAICA. 1971. Annual Report of Research Department, June 1970 - June 1971. Coconut Industry Board, Kingston,

Jamaica. p. 46-47.

COPELAND, E. B. 1914. The Coconut. Macmillan,
London. 212 pp.

CORNELIUS, J. A. 1973. Coconuts:A Review. Tropical Science 15(1):15-37.

CREENCIA, R. 1975. Perennial crop species under
coconut. National Coconut Research Symposium,
Tacloban City, November 17-19, 1975. Philippine
Council for Agriculture and Resources Research.
p. 163-169.

CSIRO. 1975. Tropical Agronomy: CSIRO Tropical
Agronomy Divisional Reports 1974 - 1975. CSIRO
Division of Tropical Agronomy, St. Lucia,
Queensland, Australia.

CUEVAS, S. E. 1975. Annual crop species under
coconuts. National Coconut Research Symposium,
Tacloban City, November 17-19, 1975. Philippine
Council for Agriculture and Resources Research.
p. 154-161.

EDACHAL, M. 1963. The papaya. Coconut Bulletin
17:34-35.

EDEN, D. R. A. 1953. The management of coconut
plantations in Western Samoa. Technical Paper
No.48, South Pacific Commission, Noumea, New
Caledonia. 32 pp.

ELLEWELA, D. C. 1956a. Report of the animal
husbandry officer. Annual Report Coconut
Research Board, Ceylon Coconut Research Institute, 1954. p. 43-44.

ELLEWELA, D. C. 1956b. Report of the animal
husbandry officer - 1955. Ceylon Coconut
Quarterly 7(1 & 2):53-55.

ELLEWELA, D. C. 1957. Report of the animal husbandry officer. Annual Report Coconut
Research Board, Ceylon Coconut Research Institute, 1955. p. 46-48.

EMPIG, L. T. 1975. Development of crop varieties for catchcropping under coconut. National Coconut Research Symposium, Tacloban City, November 17-19, 1975. Philippine Council for Agriculture and Resources Research. p. 183-187.

ERICKSEN, F. I. 1973. Grazing under coconuts at Vailele, Progress Report. WSTEC/SFP/Agriculture Department Project Report, Apia, Western Samoa. 11 pp. (mimeograph).

ERICKSEN, F. I. 1977. The effect of different light intensities on morphology, yield and nitrogen fixation rates of tropical grasses and legumes, Unpublished PhD dissertation, University of Hawaii, Honolulu, USA.

ERICKSEN, F. I. AND A. S. WHITNEY. 1977. Performance of tropical grasses and legumes under different light intensities. Proceedings Regional Seminar Pasture Research, Ministry of Agriculture and Lands, Honiara, Solomon Islands. p. 180-190.

ESPENSHADE, E. B., JR. 1964. Goode's World Atlas. Rand McNally, Chicago. 288 pp.

FAO. 1966. Coconut in Mixed Agriculture. Commodity Reports, New Series, No.1. FAO, Rome. 36 pp.

FAO. 1971. Production Yearbook. FAO, Rome. Volume 25.

FELIZARDO, B. C. 1973. Care of the fertility of soils in pasture - coconut agriculture. Paper presented at Regional Seminar - Workshop on "Pasture Production Under Coconut Palms", held at Davao City, Philippines. October 15-19, 1973. ASPAC Food and Fertilizer Technology Center, Taipei, Taiwan. 10 pp. (mimeographed).

FELIZARDO, B. C. 1975. Effect of grazing under coconut on physical and chemical properties of the soil. National Coconut Research Symposium, Tacloban City, November 17-19, 1975. Philippine Council for Agriculture and Resources Research p. 179-182.

FERDINANDEZ, D. E. F. 1968. Report of the agrostologist - 1967. Ceylon Coconut Quarterly 19(1 & 2):54-67.

FERDINANDEZ, D. E. F. 1969. Report of the agrostologist - 1968. Ceylon Coconut Quarterly 20(1 & 2):72-79.

FERDINANDEZ, D. E. F. 1970. Report of the agrostologist - 1969. Ceylon Coconut Quarterly 21(1 & 2):49-55.

FERDINANDEZ, D. E. F. 1971. Report of the Agrostology Division, 1970. Ceylon Coconut Quarterly 22:52-66.

FERDINANDEZ, D. E. F. 1972a. Effect of monospecific pasture swards on the nut yield of coconut. Ceylon Coconut Quarterly 23(1 & 2):51-53.

FERDINANDEZ, D. E. F. 1972b. Report of the Agrostology Division, 1971. Ceylon Coconut Quarterly 23:45-47.

FERDINANDEZ, D. E. F. 1973. Utilization of coconut lands for pasture development. Ceylon Coconut Planters' Review 7:14-19.

FERDINANDEZ, D. E. F. 1975. Personal communication. Coconut Research Institute of Sri Lanka.

FERGUSON, J. 1907. Coconut Planters' Manual. Eugene L. Morice, London.

FOALE, M. A. 1967. The coconut industry of the British Solomon Islands. Ceylon Coconut Quarterly 18:31-37.

FREEMAN, IAN. 1972. Personal Communication. Department of Agriculture, Honiara, Solomon Islands.

FREMOND, Y. 1966. Leguminous cover crops in coconut plantations. Oleagineux 21:437-440.

FREMOND, U., R. ZILLER AND M. DE NUCI DE LAMOTHE. 1966. The Coconut Palm. International Potash Institute, Berne, Switzerland (English edition). 227 pp.

FRIEND, D. 1977. The joint coconut research scheme, Yandina, Solomon Islands. Proceedings Regional Seminar Pasture Research, Ministry of Agriculture and Lands, Honiara, Solomon Islands.

GALL, C. 1975. Milk production from sheep and goats. World Animal Review (FAO) 13:1-8.

GALLASCH, H. 1976. Integration of cash and food cropping in the lowlands of Papua New Guinea. Proceedings, Papua New Guinea Food Crops Conference, 1975. Department of Primary Industry, Port Moresby. p. 101-115.

GANARAJAH, T. 1954. Planting systems; a summary of present-day practice in Ceylon. Ceylon Coconut Quarterly 5(2):90-93.

GHATGE, M. B. 1964. Coconut growing in Maharashtra. Coconut Bulletin 18(7):299-302.

GOMEZ, A. A. 1974. Intensification of cropping systems in Asia. In: Interaction of Agriculture With Food Science. R. Mac Intyre (Ed.). Proceedings of a Symposium in Singapore, February 1974. International Development Research Centre, Ottawa, Canada. p. 93.

GOONASEKERA, G. C. M. 1951. Rotational grazing under coconuts. Ceylon Coconut Quarterly 2(4):181-183.

GOONASEKERA, G. C. M. 1953a. Pastures under coconuts. Ceylon Coconut Quarterly 4(3 & 4):135-136.

GOONASEKERA, G. C. M. 1953b. Pastures under coconuts: the incidence of weeds. Ceylon Coconut Quarterly 4(1):9-12.

GOONASEKERA, G. C. M. 1953c. Grazing habits of indigenous cattle under coconuts. Ceylon Coconut Quarterly 4(2):73-75. (see also Tropical Agriculturist, Ceylon 110(1):25-9, 1954).

GOONASEKERA, G. C. M. 1954. Cattle under coconuts. Feeding trials with indigenous cattle. Ceylon Coconut Quarterly 5(4):227-231.

GOONASEKERA, G. C. M. 1967. The Sinhala cattle. Ceylon Coconut Planters' Review. 5(1):1-8.

GORRIE, R. M. 1950. Instructions for soil conservation practice in small-holdings and highland allotments. Ceylon Coconut Quarterly 1(4):11-16.

GOWDA, D. M. 1971. From arms to palms. Coconut Bulletin 1(12):7-8.

GRAY, S. G. 1962. Hot water seed treatment for Leucaena glauca (L.) Benth. Australian Journal of Experimental Agriculture and Animal Husbandry. 2:178-180.

GRAY, S. G. 1968. A review of research on Leucaena leucocephala. Tropical Grasslands (Australia). 2(1):19-30.

GROENEVELD, S. 1967. Landwirtschaftliche Entwicklung in Kustengebiet Ostafrikas und Geispiel der Kokospalmen. - Rindviehprojecte in der Tanga-Region in Tanzania. Munchen, Weltform Verlag. 124 p.

GUTTERIDGE, R. C. AND P. C. WHITEMAN. 1977. The regional pasture trials in the Solomon Islands. Proceedings Regional Seminar Pasture Research, Ministry of Agriculture and Lands, Honiara, Solomon Islands. p. 191-210.

GUTTERIDGE, R. C. AND P. C. WHITEMAN. 1978. Pasture species evaluation in the Solomon Islands. Tropical Grasslands 12(2):113-126.

GUTTERIDGE, R. C., P. C. WHITEMAN AND S. E. WATSON. 1976. Final Report on the Regional Programme of Pasture Species Evaluation and Soil Fertility Assessment, 1973 - 1976. Unpublished report, Solomon Islands Pasture Research Project, South Pacific Aid Programme, Australian Development Assistance Agency. 135 pp., 34 refs.

GUZMAN, MOISES R. DE, JR. 1970. Economics of Beef Production Under Coconuts in Mindanao,1968. M.S. Thesis, College of Agriculture, University of the Philippines, College, Leguna. 93 pp.

GUZMAN, MOISES R. DE, JR. 1974. Pasture and Fodder Production Under Coconuts. ASPAC Food and

Fertilizer Technology Center Extension Bulletin
No.45. Taipei, Taiwan. 29 pp.

GUZMAN, M. R., DE, JR. 1975.Pastures and pasture
management in the tropics. ASPAC Food and
Fertilizer Technology Center Extension Bulletin
No.47. Taipei, Taiwan. 28 pp.

GUZMAN, M. R., DE, JR. 1976. Cattle feeding in
the tropics. ASPAC Food and Fertilizer Technology Center Extension Bulletin No.64. Taipei,
Taiwan. 20 pp.

GUZMAN, M. R. DE, JR. AND A. V. ALLO. 1975. Pasture Production Under Coconut Palms. Food and
Fertilizer Technology Center, Taipei, Taiwan.
90 pp. 10 chapters, 26 references.

HAIGH, J. C., R. V. NORRIS, R. K. S. MURRAY, AND W. V.
D. PEIRIS. 1951. A Manual on the Weeds of
the Major Crops of Ceylon. Peradineya Manual
No.7, Department of Agriculture, Ceylon. 107 pp.

HAMPTON, R. E. 1972. Vegetables. Record of a
Coconut Industry Workshop, May, 1972. Department of Agriculture, Fiji, p. 72-74

HARTLEY, C. W. S. 1970. The Oil Palm (Elaeis
guineensis Jacq.). Longman, London. 706 pp.

HENKE, L. A. AND K. MORITA. 1954. Value of koa
haole as a feed for dairy cows. Hawaii Agricultural Experiment Station Circular No.44.
14 pp.

HILL, G. D. 1969. Grazing under coconuts in the
Morobe District. Papua and New Guinea Agricultural Journal 21(1):10-12.

HOLLAND, T. H. 1926. Soil erosion and cover crops.
Tropical Agriculturist, Ceylon 66(4-5):248-256.

HOLLAND, T. H. 1927. Cover crops and the possibilities of utilizing indigenous plants. Tropical Agriculturist, Ceylon. 68(5):263-268.

HOLM, L., D. L. PLUCKNETT, J. V. PANCHO, AND J.
HERBERGER. 1977. The World's Worst Weeds, Dis-
tribution and Biology. University Press of
Hawaii, Honolulu. 609 pp.

HOLM, L., J. V. PANCHO, J. HERBERGER, AND D. L. PLUCKNETT. 1979. Geographical Index of World Weeds. John Wiley & Co., New York.

HORNG, C. B. 1976. Some cattle diseases of tropical Asian countries. ASPAC Food and Fertilizer Technology Center Extension Bulletin No.68. Taipei, Taiwan. 20 pp.

HOSAKA, E. Y. AND J. C. RIPPERTON. 1953. Molasses grass on Hawaiian ranges. Extension Bulletin 59, University of Hawaii. 9 pp.

HOYLE, J. C. 1969. The effect of herbicides on the growth of young coconuts. Tropical Agriculture, Trinidad. 46:137-143.

HUGH, EDWIN I. 1972a. Potential for livestock production on coconut lands. Regional Seminar on Pastures and Cattle Under Coconuts. (Edwin I. Hugh, Ed.). Report of meeting, Alafua, Western Samoa, August - September 1972. South Pacific Commission, Noumea, New Caledonia. p. 69-86.

HUGH, E. I. (ed.). 1972b. Regional Seminar on Pastures and Cattle Under Coconuts. Report of meeting, Alafua, Western Samoa, August - September, 1972. South Pacific Commission, Report 215/73. Noumea, New Caledonia. 156 pp.

HUGH, E. I. 1973. Tropical Pastures for Beef Production. Information Document No.30, South Pacific Commission, Noumea, New Caledonia. 79 pp. (report of a technical meeting in May 1969 in Queensland, Australia).

HUMPHREYS, L. R. 1974. A Guide to Better Pastures for the Tropics and Subtropics. Wright, Stephenson and Co., Ltd. 15 Ascot Vale Road, Flemington, Victoria, Australia. 3031. 95 pp. 3rd edition.

IGLESIA, J. S. 1975. Coconut farming systems. National Coconut Research Symposium, Tacloban City, November 17-19, 1975. Philippine Council for Agriculture and Resources Research. p. 145-153.

IMPERIAL AGRICULTURAL BUREAU, GREAT BRITAIN. 1944.
The Provision of Animal Fodder in Tropical and
Subtropical Countries. Part 1. Bulletin 31,
Imperial Bureau of Pastures and Fodder Crops,
Aberystwyth, Great Britain. 84 pp.

JAVIER, EMIL Q. 1971. Improved varieties for pastures under coconuts. ASPAC Food and Fertilizer Technology Center Extension Bulletin No.37.
Taipei, Taiwan. 12 pp.

JAVIER, E. Q. 1975. Pastures and forages for
water buffaloes. Paper presented at ASPAC Food
and Fertilizer Technology Center seminar on
"Asiatic Water Buffalo: Its Present and Potential Value as Producer of Meat and Milk as a
Draft Animal", Khon Kaen University, Khon Kaen,
Thailand, April 1975. (mimeographed). 18 pp.

JAVIER, E. Q. 1976. Techniques of intensive for-
age production. ASPAC Food and Fertilizer
Technology Center Extension Bulletin No.81.
Taipei, Taiwan. 17 pp.

JEFFCOTT, BRUCE. 1972. Personal Communication.
Member of Parliament, Papua New Guinea.

JENSEN, B. B. 1976. Beef cattle under coconut:
tentative economic analysis basis on grazing
trials at Vailele. Economic Analysis and
Planning Division Working Paper No.9, Department
of Agriculture, Apia, Western Samoa.

JOACHIM, A. W. R. AND S. KANDIAH. 1930. The
effects of green manure and cover crops on soil
moisture. Tropical Agriculturist, Ceylon
74(1):3-9.

JOHN, C. M. 1952. Coconut Cultivation. Ernakulam,
India.

JOHNSON, H. D., A. C. RAGSDALE, I. L. BERRY, M. D.
FRANKLIN. 1963. Environmental physiology and
shelter engineering with special reference to
domestic animals. LXVI. Temperature humidity
effect including influence of acclimation in
feed and water consumption of Holstein cattle.
Research Bulletin No.846. Missouri Agricultural Experiment Stations.

JONES, R. J. 1972. The place of legumes in tropical pastures. ASPAC Food and Fertilizer Technology Center Technical Bulletin No.9. Taipei, Taiwan. 69 pp., 119 references.

JORDAN, D. AND A. A. OPOKY. 1966. The effect of selected soil covers on the establishment of cocoa. Tropical Agriculture, Trinidad. 43(2):155-166.

KANNAN, K. AND K. BHASKARAN NAMBIAR. 1973. Preliminary observations on interplanting coconut with cocoa. Coconut Bulletin 4(3):5-8.

KASASIAN, L. 1972. Report on a trip to Solomon Islands -- 11 - 25th June, 1972. Weed Research Organization Internal Report No.80. WRO, Agricultural Research Council, Begbroke Hill, Yarnton, Oxford, United Kingdon. 6 pp. (mimeographed, not a formal publication).

KASASIAN, L. AND J. SEEYAVE. 1968. Weedkillers for Caribbean Agriculture. Regional Research Centre, University of the West Indies, Jamaica and Trinidad. 44 pp.

KEE, N. S. 1968. Fertilizer problems in mixed cocoa-coconut plantings. In: Cocoa and Coconuts in Malaysia., J. W. Blencowe and P. D. Turner (Eds.), Proceedings of a Symposium in 1967, The Incorporated Society of Planters, P.O. Box 262, Kuala Lumpur, Malaysia., p. 25-31.

KIBLER, H. H. AND S. BRODY. 1950. Effects of temperature 50° to 90° F on heat production and cardiorespiratory activities in Brahman Jersey and Holstein cows. Research Bulletin No.464, Missouri Agricultural Experiment Station.

KIN, T. T. 1968. Effects of cocoa underplanting on growth and yield of coconuts. In: Cocoa and Coconuts in Malaysia., J. W. Blencowe and P. D. Turner (Eds.), Proceedings of a Symposium in 1967, The Incorporated Society of Planters, P.O. Box No.262, Kuala Lumpur, Malaysia. p. 50-56.

KINCH, D. M. AND J. C. RIPPERTON. 1962. Koa haole production and processing. Hawaii Agricultural Experiment Station Bulletin No.129. 58 pp.

306

KOTALAWALA, J. 1968. Pineapple cultivation in coconut land in the low-country wet zone. Ceylon Coconut Planters' Review 5(3):112-114.

KRISHNA MARAR, M. 1953. Intercultivation in coconut gardens - its importance. Indian Coconut Journal 4(4):131-137.

KRISHNA MARAR, M. 1961. Trial of intercultivation practices in coconut gardens. Indian Coconut Journal 14(3):87-99.

KRISHNA MARAR, M. 1964. Coconut cultivation in the Andamans and Nicobars. Coconut Bulletin 18(5): 209-217.

KRISHNA MARAR, M. AND K. PANDALAI. 1957. Influence of weather factors on the coconut crop. Indian Journal of Meteorology and Geophysics. 8:special issue.

KUTTAPPAN, M. 1971. Banana - beneficial intercrop in coconut gardens. Coconut Bulletin 1(2):2-4.

LAKSHMANACHAR, M. S. 1963. Studies on the effect of rainfall on coconut crop. A short review. Coconut Bulletin 10:370-372.

LAMBERT, M. (Ed.). 1970. Coconut Production in the South Pacific. South Pacific Commission Handbook No.6. South Pacific Commission, Noumea, New Caledonia. 117 pp.

LAMBERT, M. (Ed.). 1973. Weed Control in the South Pacific. South Pacific Commission Handbook No.10. Noumea, New Caledonia. 119 pp.

LAMBERT, M. (Ed.). 1974. South Pacific Commission Sub-regional training course on weed control in tropical pastures. Kingdom of Tonga, February 18-28, 1977. (mimeographed). 37 pp.

LEACH, B. J., S. G. REYNOLDS, AND J. H. HELLESOE. 1976. A position paper on intercropping with coconut in Western Samoa. Paper prepared for Coconut Intercropping Workshop, COCOTECH Sub-Panel on Coconut Production and Productivity, May, 1976. (mimeographed). Department of Agriculture, Western Samoa. 13 pp.

LEACH, J. R. 1968. The economics of cocoa-coconut interplanting. In: Cocoa and Coconuts in Malaysia. J. W. Blencowe and P. D. Turner (Eds.), Proceedings of a Symposium in 1967, The Incorporated Society of Planters, P.O. Box No.262, Kuala Lumpur, Malaysia. p. 103-111.

LEE, MORRIS. 1972. Personal communication. General Manager, Western Samoan Trust Estates Corporation, Apia, Western Samoa.

LEFORT, E. J. E. 1956. Economic aspects of the coconut industry in the South Pacific. South Pacific Commission Technical Paper No.92. South Pacific Commission, Noumea, New Caledonia. v + 20 pp.

LEVER, R. J. A. W. 1969. Pests of the Coconut Palm. FAO Agricultural Studies No.77, FAO, Rome. 190 pp.

LIYANAGE, D. V. 1955. Hedge planting for coconuts? Ceylon Coconut Quarterly 6(1 & 2):24-28.

LIYANAGE, D. V. 1963a. Methods of underplanting in senile coconut plantations. Ceylon Coconut Quarterly 14(3 & 4):89-94.

LIYANAGE, D. V. 1963b. Methods of underplanting in senile coconut plantations. Ceylon Coconut Planters' Review 3(4):87-90.

LOVANG, T. 1976. Weed control recommendations for some pasture weeds in Western Samoa: based mainly on spraying trials carried out May - September, 1976. Livestock and Pasture Agronomy Report Series No.5, Department of Agriculture, Apia, Western Samoa. 14 pp. (mimeographed).

LUCAS, R. J. 1968. Agricultural progress on Niue. South Pacific Bulletin First Quarter:35-39.

LUCAS, RICHARD J. 1972. Personal Communication. Department of Agriculture, Niue, South Pacific.

MAC EVOY, M. G. 1973. Economic aspects of coco/ beef. Unpublished paper presented at ASPAC Regional Seminar - Workshop on "Pasture Production Under Coconut Palms", held at Davao, Philippines on October 15-19, 1973. (mimeographed). 6 pp.

MAC EVOY, M. G. 1974. Establishment and management of pastures in coconut plantations. ASPAC Food and Fertilizer Technology Center Extension Bulletin No.38. Taipei, Taiwan. 17 pp.

MADAMBA, J. C. 1973. Recommended cattle husbandry practices in the Philippines. Unpublished paper presented at the ASPAC Regional Seminar - Workshop on "Pasture Production Under Coconut Palms", held at Davao, Philippines on October 15-19, 1973. (mimeographed). 19 pp.

MALKANI, P. G. 1951. Produce more milk. Ceylon Coconut Quarterly 2(2):83-84.

MANCIOT, R. 1968. Le cocotier aux Nouvelle - Hebrides. Premier resultats obtenus Sur la station de 'I.R.H.O. Oleagineux 23:167-174.

MANICOT, R. AND MANDRET, G. 1976. Progress realises par la Station Experimentale de Saraoutu sur l'amerlioration des paturages sous cocotiers aut Nouvelles - Herbrides. Mimeographed Report, May 1976, Station Experimentale de Saraoutu, New Hebrides. 27 pp.

MANGARAT, G. P. 1964. Companion cropping in coconut. Quezon Experiment Station, Bureau of Plant Industry, Manila, Philippines. (mimeographed).

MANIDOOL, C. 1972. Grassland farming. 3. Establishment and management of tropical pastures. ASPAC Food and Fertilizer Technology Center Extension Bulletin No.21. Taipei, Taiwan. 29 pp.

MANIDOOL, C. 1974. Quality of forage crops. ASPAC Food and Fertilizer Technology Center Extension Bulletin No.44. Taipei, Taiwan. 18 pp.

MARKOSE, T. 1973. Package of practices for coconut. Coconut Bulletin 4(2):2-9.

MATHEW, C. 1964. Water plays key role in successful coconut cultivation. Coconut Bulletin 18(1):10-12.

MATHEW, C. 1965. That button-shedding can be arrested. Coconut Bulletin 18(12):453-455.

MATHEW, C., AND A. RAMADASNAN. 1964. Effect of N, P and K nutrients on the growth of coconut seedlings. Indian Coconut Journal 17(3):114-117.

MAYALL, S. L. 1973. Survey of the copper and phosphate status of cattle in Papua New Guinea. Papua New Guinea Agricultural Journal 24(2):54-57.

MC CULLOCH, G. C. 1968. The economics of coconut mono-culture in Malaysia. In: Cocoa and Coconuts in Malaya. Proceedings of a Symposium, September 1967. (J. W. Blencowe and P. D. Turner, Editors). The Incorporated Society of Planters, P.O. Box 262, Kuala Lumpur, Malaysia. p. 96-102.

MC ILROY, R. J. 1972. An Introduction to Tropical Grassland Husbandry. Oxford University Press, Ely House, London. 147 pp.

MENON, K. P. V. AND K. M. PANDALAI. 1958. The Coconut Palm - A Monograph. Indian Central Coconut Committee, Ernakulam, India. 384 pp.

MICHAEL, K. J. 1964. Studies on the root system of the coconut palm. Indian Coconut Journal 17(2):85-92.

MIGVAR, LEO. 1965. The coconut in Micronesia. Agricultural Extension Circular No.3, Trust Territory of the Pacific Islands, Saipan, Marianna Islands. 9 pp.

MOENGANGONGO,SIAOSI. 1972. Personal Communication. Department of Agriculture, Nuku 'alofa, Kingdom of Tonga.

MOORE, R. M. (Ed.). 1970. Australian Grasslands. Australian National University Press, Canberra. 455 pp.

MUNRO, R. W. AND L. C. BROWN. 1916. A Practical Guide to Coco - Nut Planting. John Bale, Sons & Danielsson, Ltd., London. Chapter VI, "Cattle - keeping", p. 53-56.

NAIR, R. G. 1964. To keep coconut production at optimum level old palms must yield place to new. Coconut Bulletin 18(2):41-43.

NAIR, R. G. AND P. CHAMI. 1963. A survey of weeds in the fields of Central Coconut Research Station, Kasaragod. Indian Coconut Journal 17(1):40-47.

NAIR, P. K. R. AND T. P. VARGHISE. 1976. Crop diversification in coconut plantations. Indian Farming 25(11):17-19.

NAIR, P. K. R., R. VARMA, AND E. V. NELLIAT. 1974. Intercropping for enhanced profits from coconut plantations. Indian Farming 24(4):11-13.

NARAYANAN, K. M. AND I. H. LOUIS. 1965. It pays to raise intercrops in coconut gardens. Coconut Bulletin 19(1):3-7.

NATARAJAN, S. 1-75. Fertilizers are found to influence coconut wilt disease in India. World Crops 27(6):261-262.

NATHANAEL, W. R. N. 1967. The application of fertilizers to adult coconut palms in relation to theoretical concepts. Ceylon Coconut Quarterly 18:5-29.

NATIONAL SCIENCE COUNCIL, SRI LANKA. 1975. Natural Products for Sri Lanka's Future. Report of a Workshop, 2-6 June 1975. National Science Council of Sri Lanka and National Academy of Sciences of the USA. National Science Council, Colombo, Sri Lanka. 53 pp.

NELLIAT, E. V., K. V. BAVAPPA AND P. NAIR. 1974. Multi-storeyed cropping, a new dimension in multiple cropping for coconuts. World Crops 26:262-266.

NETHSINGHE, D. A. 1961. Magnesium deficiency in coconut palms. Ceylon Coconut Planters' Review 2(1 & 2):3-6.

NETHSINGHE, D. A. 1963. Manuring of young palms. Ceylon Coconut Planters' Review 3(2):41-44.

NETHSINGHE, D. A. 1966. Studies on fertilizer placement using radio isotopes. Ceylon Coconut Planters' Review 4(3):55-60.

NICHOLLS, D. F. AND PLUCKNETT, D. L. 1971.
Packet planting techniques for tropical pastures. Hawaii Farm Science 20(3):10-11.

NICHOLLS, D. F. AND PLUCKNETT, D. L. 1972.
Vegetative packets for pasture establishment.
University of Hawaii Cooperative Extension
Service Instant Information No.2.

NITIS, I. M. AND I. G. M. OKA. 1977. Feeding
behaviour of Bali cattle (Bos Banteng) raised
on improved pasture and tethered in the field
and its subsequent effect on live weight gain.
Fakultas Kedokteran Hewan dan Peternakan,
University of Udayana, Denpasar, Bali,
Indonesia. 49 pp. (Indonesian, with English
summary).

NITIS, I. M., K. RIKA, M. SUPARDJATA, K. K.NURBUDHI,
AND L. R. HUMPHREYS. 1976. Productivity of
improved pastures grazed by Bali cattle under
coconuts, a preliminary report. Dinas
Peternakan Propinsi Daerah Tingkat I Bali.
18 pp.

NORRIS, D. O. 1967. The intelligent use of
inoculants and lime pelleting for tropical
legumes. Tropical Grasslands (Australia)
1(2):107-121.

NORRIS, D. O. 1969. The importance of inoculation
in establishing tropical pasture legumes.
Technical Meeting on Tropical Pasture and Beef
Production. Brisbane, Australia. 5-17 May
1969. South Pacific Commission, Noumea, New
Caledonia. p. 105-109.

NYBERG, A. J. 1968. The Philippine coconut industry in economic perspective. The Philippine
Agriculturist 52(1):1-48.

OAKES, A. J. 1968. Leucaena leucocephala; description - culture - utilization. Advancing
Frontiers of Plant Science 20:1-114.

OHLER, J. G. 1969. Cattle under coconuts. Tropical Abstracts (The Netherlands) 24(10:639-645.

OHLER, J. G. 1972. Cattle under coconuts. Ceylon
Coconut Quarterly 23:103-107. (Reprinted from

Tropical Abstracts 24(10):639-645. Oct. 1969).

OHLER, J. G. 1974. Cattle grazing under coconuts intercropping. Journal of the Agricultural Society of Trinidad and Tobago 74(4):352-361.

ORAM, P. A. 1975. Livestock production and integration with crops in developing countries. In: Proceedings of the III World Conference On Animal Production. (R. L. Reid, Editor). Sydney University Press, Sydney, Australia. p. 309-330.

OSBORNE, H. GEORGE. 1972. Cattle production and management under coconuts. Report of Meeting, Regional Seminar on Pastures and Cattle Under Coconuts. (Edwin Hugh, Editor). South Pacific Commission, Noumea, New Caledonia. p. 141-144.

OSBORNE, H. G. 1978. Personal Communication. Consultant, Livestock Management in Coconut Plantations, 10 Carmody Road, St. Lucia, Brisbane, Queensland, Australia, 4067.

OWENS JONES, J. B. 1967. Underplanting coconut stands with cocoa on Kuala Perak Estate with special reference to planting methods and manufacturing procedures. Planter 43(3):95-98.

PALTRIDGE, T. B. 1956. Some notes on the management of coconut estates in Ceylon. Ceylon Coconut Quarterly 7(3 & 4):11-20.

PALTRIDGE, T. B. 1957. Report of the agronomist - 1956. Ceylon Coconut Quarterly 8(1 & 2):42-47.

PANCHO, J. V., M. R. VEGA AND D. L. PLUCKNETT. 1969. Common Weeds of the Philippines, Weed Science Society of the Philippines, Los Banos. 106 pp.

PANER, V. E., JR. 1975. Multiple cropping research in the Philippines. Proceedings, Cropping Systems Workshop, International Rice Research Institute, Los Banos, Philippines. p. 188-202.

PARHAM, J. W. 1955. The Grasses of Fiji. Bulletin No. 30, Department of Agriculture, Suva, Fiji. 166 pp.

PATEL, J. S. 1938. The Coconut: A Monograph, Madras, India.

PAULOSE, T. T. 1965. Grow areca nut the scientific way. Coconut Bulletin 19(5):137-145.

PEDERSEN, J. L. 1968. Maintenance and harvesting mature cocoa under coconuts. In: Cocoa and Coconuts in Malaysia, J. W. Blencowe and P. D. Turner (Eds.), Proceedings of a Symposium in 1967, The Incorporated Society of Planters, P.O. Box No.262, Kuala Lumpur, Malaysia. p. 20-24.

PERERA, M. E. 1972. Sheep breeding and management under coconut in Ceylon. Ceylon Coconut Quarterly 23:100-102.

PEREZ, C. B., JR. 1976. Fattening cattle on farm by-products. ASPAC Food and Fertilizer Technology Center Extension Bulletin No.83. Taipei, Taiwan. 11 pp.

PHILIPPINE COCONUT AUTHORITY. 1974. Annual Report 1973-74. Agricultural Research Department, Philippine Coconut Authority. p. 31.

PHILIPPINE COCONUT AUTHORITY. 1975. A Review, Coconut Research and Development Project. Agricultural Research Department, Philippine Coconut Authority. Davao. p. 20.

PHILIPPINE COUNCIL FOR AGRICULTURAL RESEARCH. 1975. The Philippines Recommends for Coconut -- 1975. Philippine Council for Agricultural Research. Los Banos, Philippines. 63 pp.

PHILIPPINE COUNCIL FOR AGRICULTURE AND RESOURCES RESEARCH. 1976a. The Philippines Recommends for Beef Cattle Production, 1976. PCARR, Los Banos, Philippines. 75 pp.

PHILIPPINE COUNCIL FOR AGRICULTURE AND RESOURCES RESEARCH. 1976b. The Philippines Recommends for Integrated Farming Systems, 1976. PCARR, Los Banos, Philippines.

PHILIPPINE COUNCIL FOR AGRICULTURE AND RESOURCES RESEARCH. 1976c. The Philippines Recommends for Pastures and Forage Crops, 1976. PCARR,

314

Los Banos, Philippines. 57 pp.

PIERIS, W. V. D. 1944. Food Production Work by the Coconut Research Scheme, Supplement to Ceylon Government Sessional Paper V of 1944.

PIERIS, W. V. D. 1945. Regeneration of coconut plantations. The Ceylon Coconut Research Scheme Bulletin No. 5. 20 pp.

PIGGOTT, C. J. 1964. Coconut Growing. Oxford University Press, London. 109 pp.

PILLAI, N. G. AND T. A. DAVIS. 1963. Exhaust of macro-nutrients by the coconut palm -- a preliminary study. Indian Coconut Journal 16(2):81-87.

PILLAI, N. G., M. V. PUSHPADAS, J. KURIAN AND S. B. LAL. 1964. The problems and prospects of coconut cultivation in Kuttanad. Coconut Bulletin 18(5):235-242.

PLUCKNETT, D. L. 1970. Productivity of tropical pastures in Hawaii. Proceedings, XI International Grassland Congress, Australia. p. A38-A49.

PLUCKNETT, D. L. 1971. Use of pelleted seed in crop and pasture establishment. University of Hawaii Cooperative Extension Service Circular No.446.

PLUCKNETT, DONALD L. 1972a. Management of natural and established pastures for cattle production under coconuts. Resource Paper for South Pacific Regional Seminar on Cattle and Pastures Under Coconuts, South Pacific Commission. Western Samoa. August - September, 1972. Unpublished manuscript. 85 pp. 124 references.

PLUCKNETT, DONALD L. 1972b. Management of natural and established pastures for cattle production under coconuts. Report of Meeting, Regional Seminar on Pastures and Cattle Under Coconuts. (Edwin I. Hugh, Editor). South Pacific Commission, Noumea, New Caledonia. p. 109-127. 42 references.

PLUCKNETT, DONALD L. 1977. Pasture research and

development in Hawaii. Proceedings Regional
Seminar Pasture Research, Ministry of Agriculture and Lands, Honiara, Solomon Islands.
p. 211-227.

PLUCKNETT, DONALD L. 1978. Integrating forage
production into small farm systems. Paper
presented at the International Conference on
"Integrated Crop and Animal Production to
optimize Resource Utilization on Small Farms
in Developing Countries", held at Bellagio,
Italy, October 18-23, 1978. The Rockefeller
Foundation, New York, New York. (typescript).
12 pp.

PLUCKNETT, D. L. AND D. F. NICHOLLS. 1972. Integration of grazing and forestry. Proceedings
of the 7th World Forestry Congress (Argentina).
7CFM/C:III/1G---(E).

PLUCKNETT, DONALD L. AND A. WHISTLER. 1977. Weedy
species of Stachytarpheta in Hawaii. Proceedings of the Sixth Asian - Pacific Weed Science
Society Conference, Jakarta, Indonesia.
Vol. I:198-203.

POMIER, M. 1967. Coconut research at Rangiroa.
Technical Paper, South Pacific Commission, No.153,
p. I - IV, 1-15.

POORTEN, E. VAN DER. 1972. Sheep under coconut.
Ceylon Coconut Planters' Review 6:133-137.

PRESTON, T. R. 1977. A strategy for cattle production in the tropics. World Animal Review
No.21. p. 11-17.

PRESTON, T. R. AND R. A. LENG. 1978. Sugar cane
as cattle feed. World Animal Review No.28.
p. 44-48.

PURSEGLOVE, J. W. 1968. Tropical Crops.
Dicotyledons 1. John Wiley and Sons, Inc.,
New York, New York. 332 pp.

RAJAPAKSE, G. 1950. Death to illuk. Ceylon Coconut Quarterly 1(4):7-9.

RAJARATNAM, D. T. AND SANTHIRASEGARAM, K. 1963.
Report of the agrostologist - 1962. Ceylon
Coconut Quarterly 14(1 & 2):35-45.

316

RAMALINGAM, N. 1961. Report of the agrostologist-1960. Ceylon Coconut Quarterly 12(1 & 2):52-73.

RANACOU, E. 1972a. Cattle grazing under coconuts. Record of a Meeting on the Coconut Industry in Fiji. Department of Agriculture, Suva, Fiji. p. 75-77.

RANACOU, E. 1972b. Pasture species under coconuts. Report of Meeting, Regional Seminar on Pastures and Cattle Under Coconuts. E. Hugh (ed.), South Pacific Commission, Noumea, New Caledonia. p. 95-100.

REBOUL, J. L. 1976. Experimentation Fourragere en Polynesie Francaise. Information Circular No.72, South Pacific Commission, Noumea, New Caledonia. 51 pp.

REPUBLIC OF THE PHILIPPINES. 1972. Annual Report, National Cooperative Pasture Resources Program. National Food and Agriculture Council, Manila. 163 pp.

REYNOLDS, S. G. 1975. Some notes on a coconut survey of the 60 acre Vailele trial area. Livestock and Pasture Agronomy Report Series No.1. Department of Agriculture, Apia, Western Samoa. (mimeographed) 26 pp.

REYNOLDS, S. G. 1976a. A summary of two years coconut data from Vailele Experimental Area: September 1974 - November 1976. Livestock and Pasture Agronomy Report Series No.6. Department of Agriculture, Apia, Western Samoa. (mimeographed). 31 pp.

REYNOLDS, S. G. 1976b. Main grass and legume species for Western Samoa. Information Sheet No.10, Department of Agriculture, Apia, Western Samoa. (mimeographed). 5 pp.

REYNOLDS, S. G. 1977a. A report on phase 3 of the cattle under coconuts grazing trial at Vailele, Western Samoa, May 1976 - February 1977. Livestock and Pasture Agronomy Report Series No.10. Department of Agriculture, Apia, Western Samoa. (mimeographed). 25 pp.

REYNOLDS, S. G. 1977b. A report on phase 3 of the

Vaea Farm Grazing Trials, 2nd July 1976 - 26th November, 1976. Livestock and Pasture Agronomy Report Series No.15. Livestock Division Department of Agriculture, Apia, Western Samoa. 14 pp.

REYNOLDS, S. G. 1977c. Expansion of the cattle industry. Alafua Agricultural Bulletin 2(1):6-12.

REYNOLDS, S. G. 1977d. Pasture research and development in Western Samoa. Proceedings Regional Seminar Pasture Research, Ministry of Agriculture and Lands, Honiara, Solomon Islands. p. 96-100a.

REYNOLDS, S. G. 1977e. The importance of water supplies in cattle production. Alafua Agricultural Bulletin 2(2):1-3.

REYNOLDS, S. G. 1977f. The utilization of coconut areas for pasture development. Working Paper No.7, 5th Regional Conference, South Pacific Commission, March 21-25, 1977. Noumea, New Caledonia. 5 pp.

REYNOLDS, S. G. 1978a. A report on the species evaluation trial under coconuts at Vailele, Western Samoa, 4 February 1975 - 22 January 1976, Livestock and Pasture Agronomy Report Series No.16, Department of Agriculture, Apia, Western Samoa.

REYNOLDS, S. G. 1978b. Cattle Under Coconuts Bibliography: A Preliminary Draft. Livestock and Pasture Agronomy Report Series No.19. Livestock Division, Department of Agriculture, Apia, Western Samoa. (mimeographed). 40 pp.

REYNOLDS, S. G. 1978c. Cattle Under Coconuts Bibliography. FAO, Rome. AGPC:Misc/54. 38 pp.

REYNOLDS, S. G. AND F. ERICKSEN. 1976. Cattle under coconuts. Information Sheet No.18, Beef-Pastures Course, Alafua College, July 5-9, 1976, Apia, Western Samoa. 11 pp.

REYNOLDS, S. G. AND T. LOVANG. 1977a. A Report on phase 2 of the Vaea Farm Grazing Trials, 5th June 1975 - 11th February 1976. Livestock and

318

Pasture Agronomy Report Series No.14. Livestock Section, Department of Agriculture, Apia, Western Samoa. 20 pp.

REYNOLDS, S. G. AND T. LOVANG. 1977b. A report on phase 2 of the cattle under coconuts grazing trial at Vailele, Western Samoa, May 1975 - March 1976. Livestock and Pasture Agronomy Report Series No.9. Department of Agriculture, Apia, Western Samoa. (mimeographed). 25 pp.

REYNOLDS, S. G., T. LOVANG, AND F. UATI. 1978. A report on phase I of the cattle under coconuts grazing trial on New Place Block at Vailele, Western Samoa, 28 May 1976 - 30 March 1977. Livestock and Pasture Agronomy Report Series No.17. Department of Agriculture, Apia, Western Samoa.

REYNOLDS, S. G. AND E. W. SCHLEICHER. 1975. A report on the cattle under coconuts grazing trial at Vailele, Western Samoa, August 1974 - February 1975. Livestock and Pasture Agronomy Report Series No.3. Department of Agriculture, Western Samoa. (mimeographed). 40 pp.

REYNOLDS, S. G. AND F. UATI. 1976. A summary of two years coconut data from Vailele Experimental Area: September 1974 - November 1976. Livestock and Pasture Agronomy Report Series No.6. Department of Agriculture, Apia, Western Samoa. 31 pp.

REYNOLDS, S. G., F. UATI, AND M. FAAMOE. 1978. A summary of coconut data from Vailele Experimental Area: September 1974 - December 1977. Livestock and Pasture Agronomy Report Series No.18. Department of Agriculture, Apia, Western Samoa.

ROBINSON, BRIAN. 1972. Comments on research needs in the region. Report of Meeting, Regional Seminar on Pastures and Cattle Under Coconuts (Edwin I. Hugh, Editor). South Pacific commission, Noumea, New Caledonia. p. 145-150.

RODRIGO, E. 1945. Fodder grass experiment (Lunuwila). Annual Report, 1945, Coconut Research Scheme, Ceylon. p. 11.

RODRIGO, P. A. AND C. P. MANAGABAT. 1964. Cacao
proves to be a paying intercrop in coconut
gardens. Coconut Bulletin 18(5):185-190.

RODRIGO, R. B., G. RAJAPAKSE AND D. C. JAYASOORIYA.
1952. Some problems of underplanting. Ceylon
Coconut Quarterly 3(3):127-131.

ROMNEY, D. H. 1964. Observations on the effect of
herbicides on young coconuts. Weed Research
4:24-30.

ROMNEY, D. H. 1965. Further experiments with
herbicides on young coconuts. Tropical Agriculture, Trinidad. 42:177-181.

ROMNEY, D. H. 1971. The dangers of using
phenoxyalkyl herbicides on coconut palms.
Ceylon Coconut Quarterly 22(3 & 4):104-106.

ROMNEY, D. H. 1972. Past studies on and present
status of lethal yellowing disease of coconuts.
PANS 18(4):386-395.

ROMNEY, D. H. 1975. Leaf spread of coconut
varieties as a measure of competition for
space. Paper read at 4th session of FAO Technical Working Party on Coconut Production,
Protection and Processing. Kingston, Jamaica.
September 14-25, 1975.

ROTAR, P. P. AND U. URATA. 1966. Some agronomic
observations in Desmodium species: seed
weights. Hawaii Agricultural Experiment
Station Technical Progress Report No.147.
University of Hawaii, Honolulu, Hawaii. 13 pp.

SAHASRANAMAN, K. N. 1963. The importance of
spacing in coconut cultivation. Coconut
Bulletin 17(1):7-11.

SAHASRANAMAN, K. N. 1964. It pays to grow groundnut in coconut gardens. Coconut Bulletin
18(4):123-130.

SAHASRANAMAN, K. N. AND MENON, K. S. 1973. Mixed
farming in coconut gardens. Coconut Bulletin
4(1):2-4.

SALGADO, M. L. M. 1947. Some problems of coconut
manuring. Papers, Ceylon Coconut Conference,

320

Colombo, Ceylon. p. 9-16.

SALGADO, M. L. M. 1951a. Cover crops for coconuts. The Ceylon Coconut Quarterly 2(2):17-20.

SALGADO, M. L. M. 1951b. Preliminary studies on the chemistry of cattle manuring on coconut estates. Tropical Agriculturist (Ceylon).

SALGADO, M. 1951c. The manuring of underplanted young palms. Ceylon Coconut Quarterly 2(4): 161-164.

SALGADO, M. L. M. 1953. Advisory work on manuring and soil management and the small-holder. Ceylon Coconut Quarterly 4(3 & 4):7-8.

SALGADO, M. 1954. Soil management on coconut estates. Ceylon Coconut Quarterly 5(3):153-158.

SALGADO, M. L. M. 1958a. Land use and soil water relations with reference to coconut cultivation. Ceylon Coconut Quarterly 9(1 & 2):12-19.

SALGADO, M. L. M. 1958b. Some pasture problems of coconut estates. Ceylon Coconut Quarterly 9(3 & 4):40-44.

SALGADO, M. L. M. 1961a. Weeds on coconut lands and their control. Ceylon Coconut Planters' Review 1(3):16-27.

SALGADO, M. L. M. 1961b. Weeds on coconut lands and their control (Continued). Ceylon Coconut Quarterly 12(4):16-19.

SAMPSON, H. C. 1928. Cover crops in tropical plantations. Tropical Agriculturist, Ceylon. 71(3):153-170.

SANDEZ POTES, A. AND E. MENO. 1972. El Cocotero. Manual de Assistincia Tecnica No.12, Institute Colombiano Agropecuario, Ministerio de Agricultura, Bogota, Colombia. 43 pp.

SANDERS, R. N. 1966. Animal selection for the tropical environment. In: Tropical Pastures, Davies, W. and C. L. Skidmore. Faber and Faber Ltd., London. p. 115-129.

SANTHIRASEGARAM, K. 1959. Report of the agrostol-
ogist - 1958. Annual Report, Ceylon Coconut
Research Institute. Government Publishing
Bureau, Colombo. p. 56-63.

SANTHIRASEGARAM, K. 1960. Observations of soil
moisture under pastures. Annual Report, Coconut Research Institute, Ceylon. 1958. p. 62.

SANTHIRASEGARAM, K. 1964. Report of the agrostologist - 1963. Ceylon Coconut Quarterly
15(1 & 2):44-49.

SANTHIRASEGARAM, K. 1965. Report of the agrostologist - 1964. Ceylon Coconut Quarterly
16(1 & 2):55-73.

SANTHIRASEGARAM, K. 1966a. Report of the agrostologist - 1965. Ceylon Coconut Quarterly
17(3 & 4):134-151.

SANTHIRASEGARAM, K. 1966b. Some Problems of Pasture Production Under Coconuts, Working Paper,
FAO Technical Working Party on Coconut Production, Protection and Processing, Colombo,
Ceylon, 1964, FAO, Regional Office for Asia and
the Far East, Bangkok, Thailand. p. 371-380.

SANTHIRASEGARAM, K. 1966c. The effect of monospecific grass swards on the yield of coconuts
in the north western province of Ceylon.
Ceylon Coconut Quarterly 17(2):73-79.

SANTHIRASEGARAM, K. 1966d. The effects of pasture
on the yields of coconuts. Journal of the
Agricultural Society of Trinidad and Tobago.
66(2):183,185-193.

SANTHIRASEGARAM, L. 19663. Utilization of the
space among coconuts for intercropping. Ceylon
Coconut Planters' Review 4(2):43-46.

SANTHIRASEGARAM, K. 1967a. Intercropping of coconuts with special reference to food production,
Ceylon Coconut Planters' Review 5(1):12-24.

SANTHIRASEGARAM, K. 1967b. Report of the agrostol-
olgist - 1966. Ceylon Coconut Quarterly
18(1 & 2):57-73.

322

SANTHIRASEGARAM, K. 1975. Effect of associated
crop of grass on the yield of coconuts. Paper
read at 4th session FAO Technical Working Party
on Coconut Production, Protection and Processing, Kingston, Jamaica. September 14-25, 1975.
11 pp.

SANTHIRASEGARAM, K. AND D. E. F. FERDINANDEZ. 1967.
Yield and competitive relationship between two
species of Brachiaria in association. Tropical
Agriculture (Trinidad) 44(3):229-234.

SANTHIRASEGARAM, K., D. E. F. FERDINANDEZ, AND
G. C. M. GOONESEKERA. 1969. Fodder grass
cultivation under coconut. Ceylon Coconut
Planters' Review 5(4):160-165.

SATYABALAN, R. K. 1972. Annual/semi-perennial
crops. Record of a Workshop on the Coconut
Industry in Fiji. Department of Agriculture,
Suva, Fiji. p. 81-84.

SCHRADER, R. S. 1950. Grass under coconuts,
Ceylon Coconut Quarterly 1(3):17-20.

SCHRADER, R. S. 1951a. Cattle for manure. Ceylon
Coconut Quarterly 2(1):15-17.

SCHRADER, R. S. 1951b. Livestock and coconut
cultivation. Ceylon Coconut Quarterly 2(2):77-
82.

SCHRADER, R. H. S. 1967. Combining the husbandry
of livestock with the cultivation of coconuts.
Papers, Coconut Conference, Colombo, Sri Lanka.
p. 17-29.

SELVADURAI, S. 1968. A Preliminary Report on the
Survey of Coconut Small-Holdings in West
Malaysia. Ministry of Agriculture and Cooperatives. Kuala Lumpur, Malaysia. 170 pp. +
appendices.

SEMPLE, ARTHUR T. 1970. Grassland Improvement.
CRC Press. Cleveland, Ohio, USA. 400 pp.

SENANAYAKE, W. 1952. Mechanized agriculture on a
coconut estate. Ceylon Coconut Quarterly
3(1):21-26.

SENARATNE, S. 1956. The Grasses of Ceylon. Government Press, Ceylon. 229 pp. plus 50 plates.

SENEWIRATNE, F. 1968. Coconut pasture project. Ceylon Coconut Planters' Review 5(2):89-91.

SESHADRI, C. A. AND P. M. SAYEED. 1953. Profitable subsidiary crops, Coconut Bulletin 7:19-21.

SETHI, B. L. 1963. Intercropping sea island cotton in coconut gardens. Coconut Bulletin 17(5):129-131.

SHANMUGAN, K. S. 1972. Fertilizing the coconut palm. World Farming 14(7):22-26.

SHELTON, H. M. 1977. Pasture research and development in Thailand. Proceedings Regional Seminar Pasture Research, Ministry of Agriculture and Lands, Honiara, Solomon Islands. p. 228-251.

SHEPHERD, R., J. R. GILBERT, AND P. G. COWLING. 1977. Aspects of cocoa cultivation under coconut on two estates in Peninsular Malaysia. The Planter 53:99-117.

SHORROCKS, V. M. 1964. Mineral Deficiences in Hevea and Associated Cover Plants. Rubber Research Institute, Kuala Lumpur, Malaysia. 76 pp.

SILVA, G. B. S. DE. 1951. Ploughing and harrowing. Ceylon Coconut Quarterly. 2(2):95-96.

SILVA, G. V. S. DE. 1961. An experiment on cattle manuring and irrigation of coconut palms. Ceylon Coconut Planters' Review 1(4):5-7.

SILVA, M. A. T. DE. 1951a. Cattle for manure. Ceylon Coconut Quarterly 2(1):15-17.

SILVA, M. A. T. DE. 1951b. Cover crops under coconuts. Ceylon Coconut Quarterly 2(1 & 2):17-22.

SILVA, M. A. T. DE. 1961. Cover crops under coconuts. Ceylon Coconut Planters' Review 11(1 & 2):17-22.

SILVA, S. DE. 1953. Cattle damage to coconuts.
Ceylon Coconut Quarterly 4(3 & 4):137-140.

SIOTA, C. M. 1973. Culture and production problems
in forage and pastures: Development of moderately to intensively managed pastures under coconuts. Philippine Council for Agriculture and
Resources Research, Crops Research Division
Workshop, Forage Pasture and Range Resources
Research Papers. University of the Philippines
at Los Banos, February 12-17, 1973. p. 41-43.

SMITH, R. W. 1968. Principles of intercropping in
coconuts with particular reference to bananas
and pastures. In: Cocoa and Coconuts in Malaya.
Proceedings, Symposium Incorporated Society of
Planters, Kuala Lumpur, Malaysia. September,
1967. p. 87-95.

SMITH, R. W. 1971. Catch cropping and intercropping. Research Department Report No.10,
Coconut Industry Board, Jamaica. p. 37-44.

SOLOMON ISLANDS JOINT COCONUT RESEARCH SCHEME.
1966-67. Annual Report. British Solomon
Islands Protectorate and Lever's Pacific
Plantations Pty. Ltd.

SOLOMON ISLANDS JOINT COCONUT RESEARCH SCHEME.
1967-68. Annual Report. British Solomon
Islands Protectorate and Lever's Pacific
Plantations Pty. Ltd.

SOLOMON ISLANDS JOINT COCONUT RESEARCH SCHEME.
1969. Annual Report. British Solomon Islands
Protectorate and Lever's Pacific Plantation
Proprietary, Ltd. 21 p.

SOLOMON ISLANDS REGIONAL SEMINAR PASTURE RESEARCH.
1977. Regional Seminar on Pasture Research and
Development in the Solomon Islands and Pacific
Region. 29 August - 6 September, 1977, Honiara.
Ministry of Agriculture and Lands, Solomon
Islands. 283 pp. (not a formal publication).

SOLOMON ISLANDS, MINISTRY OF AGRICULTURE AND LANDS.
Undated. Weeds of the Solomon Islands and Their
Control. Research Division, Ministry of Agriculture and Lands, Honiara, Solomon Islands.
108 pp.

SOUTHERN, P. J. 1967a. Sulfur deficiency in coconuts. In: Coconut Research in the South
Pacific. South Pacific Commission and Institute
de Recherches pour les Huiles et Oleagineux.
p. 1-10.

SOUTHERN, P. J. 1967b. Sulfur deficiency in coconuts, a widespread field condition in Papua and
New Guinea. 1. The field and chemical diagnosis
of sulfur deficiency. Papua and New Guinea
Agricultural Journal 19(1):17-36.

SOUTHERN, P. J. 1967c. Sulfur deficiency in coconuts, a widespread field condition in Papua and
New Guinea. 2. The effect of sulfur deficiency
on copra quality. Papua and New Guinea Agricultural Journal 19(1):38-44.

SOUTHERN, P. J. AND K. DICK. 1967. The distribution of trace elements in the leaves of the
coconut palm, and the effect of trace element
injections. In: Coconut Research in the South
Pacific. South Pacific Commission and Institute
de Recherches pour les Huiles et Oleagineaux.
p. 1-7.

SOUTH PACIFIC COMMISSION. 1972. Regional Seminar
on Pastures and Cattle Under Coconuts. (Edwin
I. Hugh, Editor). Report of Meeting, South
Pacific Commission, Noumea, New Caledonia.
156 pp.

SPENCER - SCHRADER. R. H. 1951. Livestock and
coconut cultivation. Ceylon Coconut Quarterly
2(2):77-82.

SPROAT, M. N. 1968. Important legumes and grasses
in Micronesia. Agricultural Extension Bulletin
No.8. Division of Agriculture, Trust Territory
of the Pacific Islands, Saipan, Mariana Islands.
30 pp.

STEEL, R. J. H. 1974. Growth patterns and phosphorus response of some pasture legumes sown under
coconuts in Bali. M. Agr. Sci. Thesis, University of Queensland, Brisbane, Australia.

STEEL, R. J. H. 1977. Management and weed problems
in smallholder pastures in the Solomon Islands.
Proceedings Regional Seminar Pasture Research,

Ministry of Agriculture and Lands, Honiara, Solomon Islands. p. 165-173.

STEEL, R. J. H. AND L. R. HUMPHREYS. 1974. Growth and phosphorus response of some pasture legumes sown under coconuts in Bali. Tropical Grasslands (Australia) 8(3):171-178.

SUMBAK, J. H. 1972. Use of fertilizer in coconut seedling establishment in a grassland area in New Britain. Papua New Guinea Agricultural Journal 23(3 & 4):73-79.

TAKAHASHI, M. AND J. C. RIPPERTON. 1949. Koa haole (Leucaena glauca). Its establishment, culture and utilization as a forage crop. Hawaii Agricultural Experiment Station Bulletin No.100. 56 pp.

TANCO, A. R., JR. 1973. Agricultural Sector Survey: Philippines. The General Report, January 19, 1973. Republic of the Philippines, Manila. p. 33.

TELFORD, E. A. AND N. F. CHILDERS. 1947. Tropical kudzu in Puerto Rico. Federal Experiment Station in Puerto Rico, USDA, Mayaguez, Puerto Rico. Circular No.27. 29 pp.

TEMPANY, HAROLD. 1950. The scourge of "Imperata". Ceylon Coconut Quarterly 1(4):5-6.

THAMPAN, P. K. 1970. Manuring the coconut palm. Coconut Bulletin 1(2):3-7.

THOMPSON, R. W. F. 1967. Spacing and nutrition of Cocos nucifera under atoll conditions in the Cook Islands. In: Coconut Research in the South Pacific. South Pacific Commission and Institute de Recherches pour les Huiles et Oleagineux. p. 1-3. (see also Oleagineux 23:449-451, 1968).

UNIVERSITY OF THE PHILIPPINES AT LOS BANOS. 1974. Beef Production Manual. Department of Animal Science, College of Agriculture, University of the Philippines at Los Banos. 89 pp.

VALIN, R. (DR.). 1972. Personal Communication. Service de l'Agriculture. Vila, Condominium of the New Hebrides.

327

VALLENTINE, J. F. 1965. An improved AUM for range cattle. Journal of Range Management 18(6):346-348.

VAN CHI-BONNARDEL, R. 1973. The Atlas of Africa. The Free Press (Mac Millan), New York. 355 pp.

VERBOOM, W. C. 1968. Grassland successions and associations in Pahang, Central Malaya. Tropical Agriculture, Trinidad 45(1):47-59.

VERGARA, F. P., C. B. PEREZ, JR., O. M. GATMAITAN, AND J. C. MADAMBA. 1974. Feedlot fattening of cattle in the Philippines. Farm Bulletin No.31. University of the Philippines, College of Agriculture, Laguna. 21 pp.

VERNON, A. J. 1972. Intercropping cocoa and coconuts. Record of a Coconut Industry Workshop, May 1972, Department of Agriculture, Fiji. p. 78-80.

VILLEMAN, G. 1964. The maintenance of coconut plantations. Oleagineux 19)1):26-31.

VINCENTE - CHANDLER, J., F. ABRUNA, R. CARO-COSTAS, J. FIGARELLA, S. SILVA, AND R. W. PEARSON. 1974. Intensive grassland management in the humid tropics of Puerto Rico. Bulletin No.233, University of Puerto Rico, Agricultural Experiment Station, Rio Piedras, Puerto Rico. 164 p.

WALTON, J. 1972. Cattle under coconuts. In: Cocoa and Coconuts in Malaysia (R. L. Wastie and D. A. Earp, Editors). Incorporated Society of Planters, Kuala Lumpur, Malaysia. p. 422-425.

WATSON, S. E. 1977. Results of the grazing trial under coconuts at Lingatu, Russell Islands. Proceedings Regional Seminar Pasture Research, Ministry of Agriculture and Lands, Honiara, Solomon Islands. p. 113-124a.

WEED SCIENCE SOCIETY OF AMERICA, SUBCOMMITTEE ON STANDARDIZATION OF COMMON AND BOTANICAL NAMES OF WEEDS. 1966. Standardized names of weeds. 14(4):347-386.

WEIGHTMAN, B. L. 1977. Pasture research and development in the New Hebrides. Proceedings Regional Seminar Pasture Research, Ministry of Agriculture and Lands, Honiara, Solomon Islands. p. 252-257.

WHISTLER, W. ARTHUR. 1978. Personal Communication. Department of Botany, University of Hawaii, Honolulu, Hawaii 96822.

WHITEHEAD, R. A. AND R. W. SMITH. 1968. Results of a coconut spacing trial in Jamaica. Tropical Agriculture, Trinidad. 45(2):127-152.

WHITEMAN, P. C. 1977. Pastures in plantation agriculture. Proceedings Regional Seminar Pasture Research, Ministry of Agriculture and Lands, Honiara, Solomon Islands. p. 144-153.

WHYTE, R. O., T. R. G. MOIR, AND J. P. COOPER. 1959. Grasses in Agriculture. FAO Agricultural Studies 42. FAO, Rome. 417 pp.

WHYTE, R. O., G. NILSSON - LIESSNER AND H. C. TRUMBLE. 1953. Legumes in Agriculture. Food and Agriculture Organization of the United Nations, Rome. FAO Agricultural Series No.21. 367 pp.

WIJEWARDENE, R. 1954. Mechanization problems on coconut plantations. Ceylon Coconut Quarterly. 5(4):209-217.

WIJEWARDENE, R. 1957. Coconuts under irrigation in the dry zone. Ceylon Coconut Quarterly 8(3 & 4):2-6.

YEATES, N. T. M. AND P. J. SCHMIDT. 1974. Beef Cattle Production. Butterworths, Melbourne, Australia.

YOUNGE, O. R., D. L. PLUCKNETT AND P. P. ROTAR. 1964. Culture and yield performance of Desmodium intortum and Desmodium canum in Hawaii. Hawaii Agricultural Experiment Station Technical Bulletin No.59. 22 pp plus 6 tables.

Appendixes

APPENDIX A: South Pacific Regional Seminar on Pastures and Cattle and Pastures Under Coconuts, Apia, Western Samoa, 30 August - September 13, 1972).

This meeting was sponsored by the South Pacific Commission. It was discussed in Chapter 11. The participants were:

Richard Burgess, Western Samoa
Flemming I. Ericksen, UNDP
Ian Freeman, Solomon Islands
Edwin I. Hugh, South Pacific Commission
Bruce Jeffcott, Papua New Guinea
Taalo Lauofo, American Samoa
Faatali Leavasa, Western Samoa
Morris Lee, WSTEC, Western Samoa
Richard Lucas, Niue
Asonei Marava, Western Samoa
Willie Meredith, Western Samoa
Siaosi Moengangongo, Tonga
Frank Moors, Western Samoa
Dr. George Osborne, University of Queensland, Australia
Dr. Donald Plucknett, University of Hawaii, USA
Martin Purcell, Western Samoa
Eminoni Ranacou, Fiji
Dr. K. Reddy, UNDP
Dr. Steve Reynolds, Alafua College, Western Samoa
Dr. Brian Robinson, Fiji Department of Agriculture
Dr. Rene Valin, New Hebrides
Willie Wong, WSTEC, Western Samoa

APPENDIX B: ASPAC Regional Seminar - Workshop
on Pasture Production Under Coconut Palms,
Davao City, Philippines, October 13 - 20, 1973.

This meeting was co-sponsored by the ASPAC Food
and Fertilizer Technology Center, Taipei, Taiwan,
and the National Food and Agriculture Council of the
Philippines. Three extension bulletins (de Guzman,
1974; Javier, 1974; Mac Evoy, 1974) and a book (de
Guzman and Allo, 1975) were results of the meeting.
There was no proceedings published. Papers
presented were as follows:
1. Agriculture in the Philippines, with special
   reference to coconut production. Juan T.
   Carlos.
2. Pasture species and their management. Emil
   Q. Javier.
3. Establishment and management of pastures in
   coconut plantations. Michael Mac Evoy.
4. Care of the fertility of soils in pasture-
   coconut agriculture. Bonifacio C. Felizardo.
5. Competition between the pasture-coconut plant
   community. Percival Sajise.
6. The plant-animal relationship. A. V. Allo.
7. Recommended cattle husbandry practices in
   the Philippines. Joseph C. Madamba.
8. Livestock pest and disease problems on
   cattle-coconut farms. Pablo Manolo.
9. The Development Bank of the Philippines
   (DBP) financing scheme for cattle-coconut
   farms. Mario Songco.
10. Economic aspects of cattle-coconut farming.
    Michael Mac Evoy.

The ASPAC/FFTC address is:

ASPAC Food and Fertilizer Technology Center
P.O. Box 3387
Taipei, Taiwan

APPENDIX C: Regional Seminar on Pasture
Research and Development in the Samoa Islands
and Pacific Region. Honiara, Solomon Islands,
August 29 - September 6, 1977

This seminar was sponsored by the Australian
Development Assistance Bureau, the government of the
Solomon Islands, and the University of Queensland,
Australia. The organizer was Dr. Peter Whiteman of
the University of Queensland. A proceedings of the
seminar has been published and can be obtained from
the Ministry of Agriculture and Lands, Honiara,
Solomon Islands. Papers presented that pertained to
coconut pasture systems were:
1. Some aspects of the history and future development of the cattle industry in the
   Solomon Islands. Ian B. Freeman.
2. Techniques for improving natural grasslands
   in the Pacific Region. Ian J. Partridge.
3. Pasture research and development in Papua
   New Guinea. N. Balachandran.
4. Pasture research and development in Western
   Samoa. S. G. Reynolds.
5. Pasture research and development in Fiji.
   S. Chandra.
6. Results of beef pasture research in Fiji.
   Ian J. Partridge.
7. Results of the grazing trial under coconuts
   at Lingatu, Russell Islands. S. E. Watson.
8. Pastures in plantation agriculture.
   P. C. Whiteman.
9. Soil fertility problems for pasture development in the Solomon Islands. L. D. C. Chase.
10. Management and weed problems in smallholder
    pastures in the Solomon Islands.
    R. J. H. Steel.
11. Performance of tropical forage grasses and
    legumes under different light intensities.
    Flemming Ericksen and A. Sheldon Whitney.
12. The regional pasture trials in the Solomon
    Islands. Gutteridge, R. C. and P. C. Whiteman.
13. Pasture research and development in Thailand.
    H. Max Shelton.
14. Pasture research and development in the New
    Hebrides. B. L. Weightman.

Persons knowledgeable about the coconut/
pasture/livestock system who attended this seminar
were: N. Balachandran, Ian B. Freeman, R.C.
Gutteridge, R. J. Jones, Ian J. Partridge, D. L.

Plucknett, S. G. Reynolds, Peter P. Rotar, H. M.
Shelton, R. J. H. Steel, S. E. Watson, B. L.
Weightman, and P. C. Whiteman.

APPENDIX D: Useful Weights, Measures and Equivalents.

Linear and Area Measurements

```
1 centimeter (cm)       =  0.3937 inches
100 centimeters (cm)    =  1 meter
100 centimeters (cm)    =  3.28 feet
100 centimeters (cm)    =  1.094 yards
100 centimeters (cm)    =  39.37 inches
1 inch (in)             =  25.4 millimeter = 2.54cm
1 foot (ft)             =  0.3048 m
1 yard (yd)             =  3 feet
1 yard (yd)             =  0.914 meter
1 rod                   =  5.5 yards
1 rod                   =  16.5 feet
1 rod                   =  25 links
1 mile (mi)             =  5,280 feet
1 mile (mi)             =  1,760 yards
1 mile (mi)             =  1,609.35 meters
1 mile (mi)             =  320 rods
1 mile (mi)             =  1/3 leagues
1 mile (mi)             =  8 furlongs = 80 chains
1 kilometer (km)        =  1,000 meters
1 kilometer (km)        =  0.621 miles
1 kilometer (km)        =  3,280 feet

1 square foot (sq ft)=  144 square inches
1 square foot (sq ft)=  0.111 square yards
1 square foot (sq ft)=  929 square centimeters
1 square yard (sq yd)=  9 square feet
1 square yard (sq yd)=  1,296 square inches
1 square yard (sq yd)=  0.8361 square meters
1 square meter (sq m)=  1,550 square inches
1 square meter (sq m)-  10.764 square feet
1 square meter (sq m)=  1.196 square yards
1 square rod (sq rod)=  30.25 square yards
1 square rod (sq rod)=  272.25 square feet
1 square rod (sq rod)=  25.29 square meters
1 acre (ac)             =  43,560 square feet
1 acre (ac)             =  4,840 square yards
1 acre (ac)             =  160 square rods
1 acre (ac)             =  0.4047 hectare (ha)
1 acre (ac)             =  10 square chains
1 acre (ac)             =  1 strip 8.25 feet wide
                           by 1 mile long
1 hectare (ha)          =  2.471 acres
1 hectare (ha)          =  10,000 square meters
1 square mile (sq mi)=  640 acres
1 square mile (sq mi)=  2.59 square kilometers
```

```
1 square mile (sq mi)        =  259 hectares
1 square kilometer (sq km) =  0.4059 square mi.
1 square kilometer (sq km) =  11,318,400 sq.ft.
```

Weights and Dry Measure

```
1 gram (g)                  =  1,000 milligrams (mg)
1 gram (g)                  =  0.035 ounce (oz)
1 gram (g)                  =  15.432 grains
1 ounce (oz)                =  28.35 grams
1 ounce (oz)                =  437.5 grains
1 pound (lb)                =  16 ounces
1 pound (lb)                =  453.59 grams
1 pound (lb)                =  7,000 grains
1 kilogram (kg)             =  1,000 kilograms (kgs)
1 kilogram (kg)             =  2.2046 pounds
1 cubic centimeter (cc)  =  1 milliliter
1 cubic centimeter (cc)  =  0.61 cubic inches
1 bushel (bu)               =  4 pecks
1 bushel (bu)               =  32 quarts (qts)
1 bushel (bu)               =  64 pints (pts)
1 bushel (bu)               =  35.24 liters (l)
1 bushel (bu)               =  128 cups
1 bushel (bu)               =  2,150.42 cubic in.
1 bushel (bu)               =  1.24 cubic feet
1 cubic foot (cu ft)        =  0.804 bushel
1 cubic foot (cu ft)        =  0.031 cubic yards
1 cubic foot (cu ft)        =  25.71 quarts (qt)
1 cubic foot (cu ft)        =  1,728 cubic inches
1 cubic foot (cu ft)        =  7.48 gallons
1 cubic foot (cu ft)        =  0.0283 cubic meters
1 cubic yard (cu yd)        =  27 cubic feet
1 cubic yard (cu yd)        =  202 gallons
1 cubic yard (cu yd)        =  21.72 bushel
1 cubic yard (cu yd)        =  0.7646 cubic meters
1 cubic meter (cu m)        =  1.3079 cubic yards
1 cubic meter (cu m)        =  35.314 cubic feet
1 ton (short)               =  2,000 pounds (lbs)
1 ton (long)                =  2,240 pounds (lbs)
1 metric ton (mt)           =  1,000 kilograms (kg)
1 metric ton short (mt)  =  1.1023 short ton
1 metric ton long (mt)   =  0.9842 long ton
1 short ton                 =  0.9072 metric ton
1 long ton                  =  1.0161 metric ton
```

Volume and Liquid Measures

```
1 teaspoon (tsp)            =  0.17 fluid oz.(fl oz)
1 teaspoon (tsp)            =  4.95 milliliters (ml)
1 teaspoon (tsp)            =  4.95 cubic centimeters
```

338

```
1 milliliter (ml)         = 0.0338 fluid ounce
1 milliliter (ml)         = 0.0021 pint
1 milliliter (ml)         = 0.001 liter
1 milliliter (ml)         = 0.2 teaspoon
1 cubic centimeter (cc) = 0.0338 fluid ounce
1 cubic centimeter (cc) = 0.0021 pint
1 cubic centimeter (cc) = 0.001 liter
1 cubic centimeter (cc) = 0.2 teaspoon
1 tablespoon (tbsp)       = 3 teaspoons
1 tablespoon (tbsp)       = 14.8 milliliters
1 tablespoon (tbsp)       = 14.8 cubic centimeters
1 tablespoon (tbsp)       = ½ fluid ounce
1 tablespoon (tbsp)       = 0.902 cubic inches
1 tablespoon (tbsp)       = 0.063 cup
1 fluid ounce (fl oz)     = 2 tablespoons
1 fluid ounce (fl oz)     = 6 teaspoons
1 fluid ounce (fl oz)     = 29.57 milliliters
1 fluid ounce (fl oz)     = 29.57 cubic centimeters
1 fluid ounce (fl oz)     = 0.125 cup
1 fluid ounce (fl oz)     = 1.805 cubic inches
1 cup                     = 16 tablespoons
1 cup                     = 48 teaspoons
1 cup                     = 8 fluid ounces
1 cup                     = 236.6 cubic centimeters or ml
1 cup                     = ½ pint
1 pint (pt)               = 16 fluid ounces
1 pint (pt)               = 473.2 milliliters
1 pint (pt)               = 473.2 cubic centimeters
1 pint (pt)               = 1.04 pounds
1 pint (pt)               = 28.37 cubic inches
1 pint (pt)               = 0.4732 liters
1 pint (pt)               = 32 tablespoons
1 pint (pt)               = 2 cups
1 pint (pt)               = 0.125 gallon
1 quart (qt)              = 0.25 gallon
1 quart (qt)              = 2 pints
1 quart (qt)              = 4 cups
1 quart (qt)              = 32 fluid ounces
1 quart (qt)              = 64 tablespoons
1 quart (qt)              = 57.75 cubic inches
1 quart (qt)              = 946.3 milliliters
1 quart (qt)              = 946.3 cubic centimeters
1 liter (l)               = 1,000 milliliters
1 liter (l)               = 1,000 cubic centimeters
1 liter (l)               = 0.264 gallon (US)
1 liter (l)               = 2.164 pint
1 liter (l)               = 33.814 fluid ounces
1 liter (l)               = 61.02 cubic inches
1 liter (l)               = 0.035 cubic foot
```

1 U.S. gallon (gal)	=	3.785 liters
1 U.S. gallon (gal)	=	4 quarts
1 U.S. gallon (gal)	=	8 pints
1 U.S. gallon (gal)	=	128 fluid ounces
1 U.S. gallon (gal)	=	16 cups
1 U.S. gallon (gal)	=	0.134 cubic feet
1 U.S. gallon (gal)	=	0.83 British gallon
1 U.S. gallon (gal)	=	3,785.3 milliliters
1 U.S. gallon (gal)	=	3,785.3 cubic centimeters
1 U.S. gallon (gal)	=	231 cubic inches
1 U.S. gallon (gal)	=	8.3358 pound water
1 cubic foot of water	=	62.2 pounds
1 pound of water	=	27.68 cubic inches
1 pound of water	=	0.016 cubic feet
1 acre-in (acre-in) of water	=	27,154 gallons
1 acre-in (acre-in) of water	=	624.23 gal per 100 sq ft

Miscellaneous Equivalents

1 gram per liter or kilogram	= 0.83 lb per U.S. gal
1 gram per liter or kilogram	= 1,000 parts per million
1 part per million (ppm)	= 1 mg per liter or kg
1 part per million (ppm)	= 0.0001 percent (%)
1 part per million (ppm)	= 0.013 oz by weight
1 part per million (ppm)	= 0.379 g in 100 gallons
1 percent (%)	= 10,000 parts per million
1 percent (%)	= 10 grams per liter
1 percent (%)	= 1.28 oz by wt per gal
1 percent (%)	= 8.336 lb per 100 gal

APPENDIX E: Institutions and Persons that are
knowledgeable about, or that have special
interest in cattle or pasture under coconut.

Institutions

ASPAC/FFTC
5th Floor 14
Wenchow Street
Taipei, Taiwan

Central Plantation Crops Research Institute
Kasaragod, Kerala
India

Coconut Industry Board
P.O. Box 204
Kingston 10, Jamaica

Coconut Research Institute
Lunuwila, Sri Lanka

Department of Agriculture
Apia, Western Samoa

Indian Coconut Research Station
Kayankulam
Kerala State
India

IRHO
Station Experimentale de Saraoutou
B. P 89
Santo, New Hebrides

Ministry of Agriculture, Fisheries and Forest
Suva, Fiji

Philippine Coconut Authority
P.O. Box 295
Davao City, Philippines

South Pacific Commission
P.O. Box D.5
Noumea, New Caledonia

Western Samoa Trust Estates Corporation
Apia, Western Samoa

Persons

Antoine, A.
Department of Agriculture Economics and Farm
  Management
University of West Indies
St. Augustine, Trinidad and Tobago

Boonklinkajorn, P.
Agricultural Research Department
Applied Scientific Research Corporation of
  Thailand
Bangkhen, Bangkok 9
Thailand

Ericksen, Flemming I., Dr.
Nylokke Vej 58, 8340, Malling, Denmark

Ferdinandez, D. E. F.
Coconut Research Institute
Lunuwila, Sri Lanka

Freeman, I.B.
Principal Veterinary Officer
Ministry of Agriculture and Lands
P.O. Box G.11
Honiara, Solomon Islands

Gunawardene, K. J.
Industrial Crops Group
Plant Production and Protection Division
Food and Agriculture Organization of the
  United Nations
Via delle Terme di Caracella
00100 - Rome, Italy

Guzman, M. R. de
5th Floor 14, ASPAC/FFTC
Wenchow Street
Taipei, Taiwan

Javier, Emil, Dr.
Director, Institute for Plant Breeding
University of the Philippines at Los Banos
College, Laguna, Philippines

Madamba, Joseph, Dr.
ICLARAM
Manila, Philippines

Nitis, I. M., Dr.
Universities Udayana
Denpasar, Bali,
Indonesia

'Ofa Fa'anunu, H.
Department of Agriculture
Nuku'alofa, Tonga

Ohler, J. G.
Agricultural and Research Department
Royal Tropical Institute
Amsterdam, The Netherlands

Osborne, G., Dr.
Veterinary School
University of Queensland
St. Lucia, Qd. 4067
Australia

Plucknett, Donald L., Dr.
Professor of Agronomy
College of Tropical Agriculture
University of Hawaii
Honolulu, Hawaii 96822, USA

Reynolds, S. G., Dr.
FAO Livestock Project
URT/78/028, P.O. Box 159
Zanzibar, Tanzania

Riveros, Fernando, Dr.
Tropical Pasture Improvement Specialist
Food and Agriculture Organization of the
  United Nations
Via delle Terme di Caracella
00100 - Rome, Italy

Santhirasegaram, K., Dr.
Tropical Pasture Agronomist
DOM 71/516
Apartado 1424
Santo Domingo, Dominican Republic

Steel, R. J. H.
Research Officer, Pasture Management
Ministry of Agriculture and Lands
P.O. Box G.11
Honiara, Solomon Islands

Watson, S.
Department of Agriculture
University of Queensland
St. Lucia, Qd. 4067
Australia

Weightman, B.
Department of Agriculture
Port Vila, New Hebrides

Whiteman, P. C., Dr.
Senior Lecturer in Tropical Agronomy
University of Queensland
St. Lucia, Qd. 4067
Australia

Index

abaca   104
Africa   10, 16, 68, 81, 200, 208, 212, 216, 219,
    222, 224, 227, 228, 283 - 284
African star grass   (see Cynodon plectostachyus)
Ageratum conyzoides   51, 57, 132
Ageratum houstonianum   57
Agusan Province, Phil.   131
Ahmad Bavappa, K. V.   279
Alabang X   (see Dicanthium aristatum)
Alocasia macrorrhiza   111
alyce clover   (see Alysicarpus vaginalis)
Alysicarpus vaginalis   117, 126
American Samoa   275 - 276
Amorphophallus campanulatus   101
Anacardium occidentale   6, 107
Ananas comosus   105
Anderson, G. D.   284
Angleton grass   (see Dicanthium aristatum)
animal management, research needs   286
animal manure(s)   23, 37
animal unit equivalent(s)   247, 248
annual field crop yields   106
Arachis hypogea (peanut, ground nut)   100, 106
Areca catechu   105
arecanut   105
Argemone mexicana (mexican poppy)   263 - 264
arrowroot   106
artifical insemination   255 - 256
Asclepias curassavica   51, 58
Asia   1, (land planted to coconut)-7, 6, 7, 25,
    61, 68, 211, 221, 277 - 280
ASPAC   260
ASPAC Food and Fertilizer Technology Center
    (FFTC)   276, 280
ASPAC training course on Pasture production under
    coconuts.   276 - 277
atoll(s)   31, 89
atrazine   82
Australia   176, 203, 217, 271, 276, 284
Australian Development Assistance Bureau   277
Axonopus compressus (carpetgrass)   12, 27, 47, 48,
    51, 115, 117, 118, 119, 120, 134
Bali   12
balsam pear   70
banana   104, 112
batiki blue grass (see Ischaemum aristatum)
beef animals   7
beef production under coconut, economics of   9
beggar - ticks   59
Belize   11

Benin   10
Bennett, Dr. Steve   259
bermudagrass   117
Bidens pilosa   48, 51, 59
billy goat weed   57
bitter melon   70
black gram   93
black pepper   102
Blechum pyramidatum   132
bloodflower milkweed   58
Boehmeria nivea   106
Borreria spp.   134
Bos banteng (Bali cattle)   12, 249
Bos indicus   119, 250, 272
Bos taurus   249 - 251
Brachiaria brizantha (palisade grass)   15, 43, 71,
    82, 147, 148, 149, 153, 160 - 175, 183, 186,
    208 - 209, 210, 216, 242, 248
Brachiaria decumbens   134, 168, 172, 173, 175,
    186, 210 - 211, 216
Brachiaria dictyoneura   (see Brachiaria humidicola)
Brachiaria humidicola   134, 176, 186, 211
Brachiaria miliiformis (cori grass)   13, 82, 138,
    139, 153, 159, 160, 161, 162, 163, 164, 167,
    shade tolerance - 168, 170, 171, 172, 186, 211-213,
    216, 248, 277
Brachiaria mutica (para grass)   15, 47, 48, 51, 82,
    117, 134, 146, 152, 170, 174, 175, 182, 212,
    214-215, 216
Brachiaria ruziziensis (ruzi grass)   170, 185, 186,
    216-217
Brazil   31, 205, 222, 225, 271
breeding stock   271-272
British West Indies   11
broom weed   75
browse system of grazing   186
buffalo grass   (see Stenotaphrum secondatum)
Burma   8
cacao   13, 93, 103, 107, 111, 189, 282
cadangcadang disease   26
Cajanus cajan (pigeon pea)   144
Calo, Dr. Lito   8
Calocarpum sapota   107
calopo   89, 117
Calopogonium mucunoides (calopo)   89, 117, 134,
    144, 172, 190, 193, 194, 231
calving   251-252
Canavalia seriacea   89
Capsicum annuum   102
Caribbean   17, 26, 104, 196

348

coconut(s)
  disease(s)   3, 25
  drought, drought effects   3, 23
  economic life   21
  effect on intercrops   94-95
  effect of grazing and pastures on yields   160
  fertilization of palms   31, 32, 33, 35
  fertilizer placement   37
  fire, effect on   19, 24
  flowers, flowering   20
  fruit set   21
  grouped plantings   113
  growing conditions   19
  harvesting, nut collection   26, 27, 28, 29, 30
  hedge planting   113
  hurricane damage   16
  husk burial   23
  insects   3
  iron nutrition   34
  irrigated   33
  lightning damage   3
  magnesium nutrition   32, 34
  manganese nutrition   34
  "manure circle"   37, 39
  marginal lands   23
  mature palms, nutrition   32, 34-36
  minor elements   36
  mixed cropping   3
  mulching   24
  nitrogen nutrition   31, 34, 35
  nutrient removal in yield   34
  nutrition, nutritional requirements   31
  oil   1, 5
  on small farms   1
  palm spacing   15, 38, 39, 40, 44
  pests   25
  phosphorus nutrition   31, 34, 35
  plantations, plantation crop   1, 12, 15, 16
  potassium nutrition   31, 34, 35
  producing countries   2, 10, 11
  rainfall, and   21, 23
  rain fed   33
  rat damage   25
  rehabilitation of groves   5, 44
  root system   19, 20, 135
  seedling nursery   22
  shading by   95, 96, 97, 98
  soil acidity   23
  soil management   24
  soils   23

coconut(s)
  sole planting   5
  strip planting   113
  spacing of palms   113
  sulfur nutrition   36
  tillage under palms   19, 24, 136, 137, 138, 139, 140
  underplanting   5, 32, 38, 41, 42, 43, 44
  understory cover management   46
  water conservation   24
  waterlogged soils   23
  weed control   7, 19, 47, 50, 81-87
  weeds of   51-56
  world crop area   10, 11
  young palms   31, 33, 42
Coconut Industry Board, Jamaica   18, 283
coconut pastures, desirable characteristics   157
coconut/pasture/beef systems
  meat production on   16
coconut press cake (copra meal, poonac)   257
Coconut Research Institute of Ceylon   13, 49, 129, 170, 227, 277
"Coconut Triangle"   13, 251, 277
coconut yield(s)   30, 31
    yield goal   31
    effect of intensity of grazing and level of fertilizer on   164
    effect on N and K pastures on   163
Cocos nucifera   1
cocoyam   (see tanier)
Coffea spp.   103
coffee   93, 103, 109, 111
Colocasia esculenta   101, 106
Colombia   17
Commelina species   117
commercial seed sources   271
communal village coconut lands   15, 16, 266-268
Comoro Islands   17
condiments, intercropping with   102
continuous grazing   128, 155, 166, 236, 240-241
Cook Islands   10
copra   1, 5, 26, 30, 35, 39
copra cutting and drying   26
coral soils   34
Corchorus spp.   106
cori grass   13
coriander   102
Coriandrum sativum   102
Cotabato Province, Phil.   131
cotton   103

351

353

Lowlands Agricultural Experiment Station, Papua
  New Guinea   282
Lucas, Richard   16, 282
Mac Evoy, Michael   279
Macroptilium atropurpureum (siratro)   89, 145, 168,
  173, 176, 179, 185, 187, 203-206, 218, 227
Macroptilium lathyroides (phasey bean)   89, 145,
  175, 209, 215, 218, 226, 227
Madamba, Joseph C., Dr.   279
Magat, S. S.   279
maize   93, 99, 106
Makasiale, James   282
Malagasy Republic   10
Malaita Island, Solomon Islands   182
Malayan Dwarf   87
Malaysia   10, 12, 51, 54, 55, 101, 103, 104, 123,
  189, 274, 275
Mangifera indica   105
mango   105
Manihot esculenta   100, 106
Manolo, Pablo, Dr.   279
manures   23, 37
Maranta arundinacea   106
marketing, research needs   286
mating   251-253
MCPB   82
Melanesia   25
Melinis minutiflorus (molasses grass)   221-224
Menon, K. S.   279
Mexico   1, 11, 17
Micronesia   25, 89
mikania   68, 117
Mikania cordata   48, 53, 68, 115, 117, 119, 122,
  134
Mikania micrantha   53, 68, 132
mile-a-minute   68
milk cow(s)   7
millets   99
Mimosa invisa   48, 53, 69
Mimosa pudica   27, 48, 53, 69, 82, 89, 115, 118,
  134, 266
Mimosa spp.   83
mimosine   203
Mindanao   8, 129, 130, 131, 138, 182, 276, 278
mineral supplements   260-261
Misamis Occidental Province, Phil.   131
Misamis Oriental Province, Phil.   131
mission lands   15
mixed farming   3, 5, 6
Moengangango, Siaosi   16

356

molasses grass   (see _Melinis minutiflorus_)
Moluccas   12
Momordica charantia   53, 70
monuron   82
Moors, Frank   15
Mozambique   10, 17, 31, 113
MSMA   86
Mucuna stizolobium   89
Mulifanua Plantation   43
mung bean, mungo   100, 106
Musa spp.   104
Musa textilis   104
Nair, P. K. R., Dr.   279
napier grass   (see _Pennisetum purpureum_)
National Academy of Sciences (USA)   278
National Food and Agriculture Council, Philippines
  276
National Science Council of Sri Lanka   278
natural pastures
  carrying capacity of   117, 128, 130, 133
  dry matter yield   132
  important grasses and legumes   116-127
  palm density and carrying capacity   131
  research   129, 131
  research needs   285-286
  use of livestock to control weeds   114
Navua sedge   63
Nelliat, E. V.   279
Nephrolepis biserrata   53, 71, 83, 134
Nephrolepis exaltata   53, 71, 83
Nephrolepis ferns   50, 53, 71, 83, 115
Nephrolepis hirsutula   53, 71
nettle-leaved vervain   78
New Britain   282
New Caledonia   276
New Hebrides, N. H. Condominium   10, 15, 16, 31,
  51, 52, 53, 54, 55, 117, 123, 126, 130, 172, 176,
  185, 248, 275
Nicaragua   17
Nigeria   6, 10
night-penning of animals   244, 247
nitrogen, N   31, 33, 94
Niue   16, 187, 268, 275, 276, 282-283
nurseries (for pasture grasses and legumes)   147,
  148
nutgrass   64
nutritional priorities for classes of cattle   240
nutritional requirements of pasture plants   157,
  158
Oceania   1, 5, 14, 68, 277

357

358

pasture establishment
  low density plantings   146
  planting in circles or strips   140
  seed beds   141
  tillage   139
peanut   (see groundnut)
Pennisetum polystachyon   15
Pennisetum purpureum (napier grass)   152, 153, 179,
  180, 208, 228, 244, 273
pepper, black   (see Piper nigrum)
Pethiyagoda, U., Dr.   278
Phaseolus atropurpureus   203
Phaseolus aureus   100
Phaseolus mungo   100, 106
phasey bean   (see Macroptilium lathyroides)
Philippine Bureau of Plant Industry   279
Philippine Coconut Authority (PCA)   278
Philippine Coconut Research Institute (PHILCORIN --
  see Philippine Coconut Authority)
Philippine Council for Agriculture and Resources
  Research   278
Philippines   1, 10, 22, 26, 31, 45, 51, 52, 53,
  54, 55, 56, 95, 99, 100, 101, 102, 103, 104, 105,
  106, 107, 108, 109, 110, 129, 130, 138, 139, 175,
  181, 182, 218, 245, 251, 274, 275, 276, 278-279
phosphorus, P   31, 33
photosensitization caused by plants   67
Phyllanthus urinaria   132
pine forest grazing, USA   235
Piper nigrum   102, 123
pineapple   93, 105, 107, 108, 109
plantain   93, 104, 110, 111
Plucknett, Donald L.   284
poisonous plants   58, 261, 263-264
Polynesia   25, 31
Polynesia, French   10
potassium, K   31, 33, 94
prickly solanum   77
Principe   16
Pseudelephantopus spicatus   54, 73, 132
Psidium guajava   48, 54, 74, 83, 107
Pueraria phaseoloides (puero, tropical kudzu)   145,
  172, 173, 174, 187, 189-192, 194, 215, 224, 231,
  236, 283
puero   (see Pueraria phaseoloides)
Puerto Rico   11, 218
pulses   (see grain legumes)
purple nutsedge   64
"quincunx"   38
ragi millet   99, 107

soil
  compaction by animals   7, 273
  erosion, prevention of   19, 24
  fertility   7
  moisture-holding capacity   7
  structure   7
soilage   (see zero grazing)
Solanum torvum   48, 55, 77
Solomon Islands   10, 15, 31, 52, 53, 54, 55, 63,
  83, 87, 114, 117, 119, 123, 130, 131, 134, 139,
  170, 172, 173, 176, 182, 183, 222, 234, 248, 275,
  276, 277, 281-282, 284, 287
Solomon Islands Ministry of Agriculture and Lands
  83, 281-282
Soothipan, Dr. Arkoon   11
sorghum   93, 99, 107, (forage) 152
Sorghum bicolor   99
Sorghum halepense (Johnson grass)   263
sour paspalum (see also Paspalum conjugatum)   117,
  121, 122, 123, 134
South Africa   271
South America   11, 17, 190, 198, 202, 227
South Pacific   83-86, 114, 123, 128, 176, 182, 187,
  268, 275
South Pacific Commission   14, 275, 276, 280
South Pacific Regional Seminar on Raising Cattle
  Under Coconuts   14, 275-276
Southeast Asia   8, 12, 25, 275
southern sida   75
soybean   106
Spanish needle   59
specialty crops   102
spices, intercropping with   102
Sri Lanka   1, 10, 13, 22, 28, 29, 30, 31, 32, 34,
  35, 39, 42, 49, 52, 53, 54, 82, 83, 87, 88, 92,
  93, 99, 100, 102, 103, 104, 105, 109, 110, 111,
  117, 123, 126, 130, 138, 139, 153, 160, 171, 172,
  173, 177, 212, 241, 242, 244-247, 251, 257, 270,
  272, 275, 277-278
Stachytarpheta spp.   48, 55
Stachytarpheta cayennensis   55, 78
Stachytarpheta indica   78
Stachytarpheta jamaicensis   55, 78, 79
Stachytarpheta urticaefolia   55, 78, 79
St. Augustine grass   (see Stenotaphrum secondatum)
Stenotaphrum secondatum (St. Augustine grass,
  buffalo grass)   117, 126, 185
stinking cassia   60
stocking rate(s)   153, 154, 167, 247-249

361

stylo   (see <u>Stylosanthes guyanensis</u>)
<u>Stylosanthes guyanensis</u> (stylo)   134, 145, 168, 170, 205, 207, 215, 243
subsidiary crops   90-113
sugarcane   105, 110, 186
Sulawesi, North   12
Sumatra   12
sunflower   106
sunn hemp   107
supplemental feeds   257-260
  coconut cake   257
  crop residues   257
  cull fruits, root crops   257
  fish meal   257
  maize, sorghum stover   257
  pineapple leaves, bran, peels   257
  rice hulls   257
  rice straw   259
  sugarcane tops   257
  sweet potato vines   257
  urea-molasses   257-259
supplementation (feed) on pasture   156, 256-261
Surinam   11
sweet potato   101, 110
sword fern   71
synedrella   80
<u>Synedrella</u> <u>nodiflora</u>   48, 56, 80, 134
<u>Tamil Nadu, India</u>   26
Tanga Livestock Experiment Station   267, 283-284
Tanga, Tanzania   266-268, 283-284
tanier   101
Tanzania   10, 17, 83, 87, 104, 105, 266-268, 275, 283-284
taro   101, 106
TCA   82
<u>Tectona grandis</u>   107
<u>tethering</u> animals to palms   16, 244-146
Thailand   10, 31, 172, 175, 275, 280
Thailand, Applied Scientific Research Corporation   280
Thailand Department of Agriculture   280
Thailand Livestock Department   280
<u>Theobroma cacao</u>   13, 103
<u>three-tiered</u> system, intercropping   107
Togo   10
Tonga, Kingdom of   10, 16, 83, 87, 117
tree crops, tree-like crops   103
<u>Tribulus</u> <u>cistoides</u>   263
<u>Tribulus</u> <u>terrestris</u> (puncture vine)   263
<u>Trigonella</u> <u>foenumgraecum</u>   102

Wong, Willie   281
WSTEC   (see Western Samoa Trust Estates
  Corporation)
Xanthium spinosum (cocklebur)   263
Xanthium strumarium (cocklebur)   263
Xanthosoma sagittifolium   101
yellow guava   74
yams   101
yautia   (see tanier)
yields, annual field crops   106
Zamboanga del Norte Province, Phil.   131
Zamboanga del Sur Province, Phil.   131
Zanzibar   105
Zea mays   99
Zebu cattle   (see Bos indicus)
zero grazing   155, 156, 179, 243-244, 273
Zingiber officinale   102

9 780367 167554